THE GREENING OF OZ

SUSTAINABLE ARCHITECTURE IN THE WAKE OF A TORNADO

by Robert Fraga

Wasteland Press

www.wastelandpress.net
Shelbyville, KY USA

The Greening of Oz
by Robert Fraga

Copyright © 2012 Robert Fraga
ALL RIGHTS RESERVED

Second Printing – June 2012
ISBN: 978-1-60047-715-7
Library of Congress Control Number: 2012937404
Grateful acknowledgement is made for permission to quote
from copyrighted works as indicated in the footnotes.
Jacket serigraph: Downtown Survivor, by Justin Marable

Except for brief text quoted and appropriately cited in other works,
no part of this book may be reproduced in any form, by
photocopying or by electronic or mechanical means, including
information storage or retrieval systems, without permission in
writing from the copyright owner/author.

Printed in the U.S.A.

0   1   2   3   4   5

Contents

THE CONTEXT	3
A LITTLE HISTORY	17
THE NIGHT IT HAPPENED	29
VOLUNTEERS	49
KATRINA'S GHOST	65
GOING GREEN	87
MEDIA MATTERS	115
THE FRUITS OF GREEN LABOR	129
TRAGEDY TO TRIUMPH	149
REALITY CHECK	165
AFTERMATH OF COMPROMISE	183
CAN THE CENTER HOLD?	201
ACKNOWLEDGMENTS	225

THE CONTEXT

In early May, 2007, a killer wind wiped out a small town in south central Kansas. The name of the town: Greensburg. The wind which blew it off the map was the biggest tornado to touch down in this country since the adoption three months earlier of the so-called Enhanced Fujita Scale.

Why write a book about what happened in Greensburg? Towns, after all, get hit by tornadoes every year. There is even a part of the Midwest called Tornado Alley. Greensburg lies smack dab in the middle of it. It's the way the town chose to rebuild that makes it unique. These are still early days. Only five years have elapsed since the tornado; but the Greensburg experiment—and it *is* an experiment—needs to be followed, not just by its own tornado-prone region of the country, but by the country as a whole. And followed attentively. This village, sequestered on the Great Plains of America, may help point the way to the future, the way human beings can survive not only increasingly violent natural disasters—a result of global climate change—but the depletion of oil and gas reserves, a growing scarcity of all but sustainable materials, and pollution so pervasive that it is poisoning our environment.

What was it that Greensburg decided to do? On December 17, 2007, the City Council mandated that all public buildings with a footprint exceeding 4,000 square feet be built to meet LEED platinum certification. It encouraged residents, reconstructing their homes and businesses, to do likewise.

We need to backpedal here. What's LEED? What's platinum certification? To answer those questions, it is necessary to introduce an organization—the U.S. Green Building Council (USGBC). Founded in Washington, D.C., in 1993, it encompasses more than 6,000 entities drawn from the ranks of government, non-profits, academia, and real estate developers. The USGBC gives as its mission:

> To transform the way buildings and communities are designed, built, and operated, enabling an environmentally and socially responsible, healthy, and prosperous environment that improves the quality of life.

To implement its goals, the USGBC launched a rating program in 2000. It was called—and is still called—the Leadership in Energy and Environmental Design (LEED). It awards points toward certification for certain materials and design features. This was later extended to certify the renovation of existing buildings. There are four levels: basic certification, silver, gold, and platinum. The point system to earn certification changed in the five years which followed the Greensburg tornado. To attain platinum, for example, it was necessary in 2007 to acquire at least fifty-two points. That changed in 2009 to eighty. The changes included three "major enhancements": harmonization, credit weightings, and regionalization.

By voting to obtain platinum certification, the Greensburg City Council was shooting for the green stratosphere. No other small town in this country had yet attempted to do what Greensburg committed itself to doing on December 17. There are (and have been for some time) cities in the US which require LEED certification for their public buildings. The December 17 vote was not unprecedented in that respect. But the Greensburg vote was the centerpiece of a smorgasbord of green moves which the town opted to make. That venture was unique in this country.[1]

In this century, going green has entered the architectural mainstream. LEED certification has become a recognized standard. Yet, as you might expect, there has been criticism of the system. Canny developers have taken advantage of loopholes in the certification process, sometimes to the detriment of sustainable construction. In one lab makeover, for example, counter tops were made of a recycled paper fiber product in order to garner a LEED point. But the product manufacturer warned that it was not designed to be used this way. Result: The counter tops soon had to be replaced.

Some architects maintain that LEED certification has more to do with good PR than with sustainability. And pursuing certification does not come cheap. Consultants able to wade through the necessary paperwork cost big bucks. The bare-bones certification fees themselves can run in excess of $10,000, an issue which impacted rebuilding in Greensburg. At least one critic has wondered, wouldn't it be better to sink the money into making a building truly green rather than certifying it to be that politically correct color? When an architecture firm or a branch of the USGBC volunteers to do this work *pro bono publico*, it is a generous offer. Nonetheless the

---

[1] Greening efforts in other countries will be touched upon elsewhere.

fact remains: LEED certification, or its equivalent, is entrenched in the environmental firmament of first-world countries.

To be fair to the USGBC, it was well known that LEED was "clumsy and limited"—to quote a former board member of USGBC, Bob Berkebile, an architect who plays a major role in the Greensburg story. "Many wanted to wait until it could be put on more scientific footing," he said, "but more wanted to get something out quickly."[2] Berkebile continued:

> What was shocking was that many agencies and cities so quickly embraced it as their tool, not realizing that it was not regional, did not do life-cycle analysis, and was focused on corporate buildings.

The 2009 revision of LEED certification was meant to correct its shortcomings.

There are other rating systems that compete with LEED—Canada's web-based Green Globes is one such—but through strength of numbers if nothing else, LEED is not likely to wither on the vine and blow away any time soon. Vivian Manasc, a founding member of Canada's Green Building Council, has said that:

> No other rating system is as broadly based in the marketplace as is LEED. With USGBC's 4000-plus members getting to vote on what is in the rating system, LEED has large public input. It's easy to write an elegant system as long as you don't have to deal with the messiness of the marketplace.[3]

The Greensburg City Council's decision to go green so uncompromisingly was doubly amazing in light of the temperament of the community. Greensburgers are no Chardonnay-sipping and Brie-nibbling liberals. One businessman even told a visiting filmmaker that he thought global warming was bullshit. Kiowa County—of which Greensburg is the county seat—is one of the reddest counties in the country, voting 80 percent plus for the Republican candidate for President in each of the elections of 2000, 2004, and 2008.[4] Greensburg could never be called the Berkeley of the Plains. The tension between the progressive thrust to go green and the dyed-in-the-wool conservatism of the town is one of the issues to be examined in this book.

---

[2] Nancy B. Solomon, "How is LEED Faring After Five Years in Use?" *Architectural Record*, June 2005.

[3] *Ibid.*

[4] George Bush amassed 1262 votes to Al Gore's 294 in 2000; 1275 against John Kerry's 256 in 2004; John McCain got 912 votes to Barack Obama's 200 in 2008. Source: Office of Kansas Secretary of State.

Why *did* the town go green? Two reasons:

1. Common sense. People understood that their investment in efficient HVAC units, in super-strong insulation, in wind farms and water conservation would pay off over time. This appealed to the practicality for which Midwesterners are renowned.

2. Religious commitment. Be good stewards of the earth, Scripture admonishes. *The Green Bible*, published in 1989 by the National Council of Churches, makes explicit—by color coding—what the Good Book has to say about this. A foreword, provided by Archbishop Desmond Tutu, claims that "God . . . created humans to be his viceroys and to act 'compassionately and gently' toward all forms of life."

"It's the necessary thing to do," agreed Marvin George, Greensburg's Baptist minister, a tattooed bear of a man. "God gave us responsibility [for the earth], and this is the way to go." Such an assertion packs a punch in a place like Greensburg, where so many people feel naked unless they're wearing their religion on their sleeves.

How well-founded were the environmental concerns that underlay the decision to go green in Greensburg? There are more books than you can shake a stick at about various aspects of the environmental challenges we face: global warming, peak oil, ground, air and water pollution, carbon footprints, population explosion. A comprehensive review of the literature lies beyond the scope of this work. But majority sentiment in the scientific community is that change is coming. The question is whether the change will involve a soft landing or a crash.

The organization responsible for measuring that change is a group established in 1988 and called the Intergovernmental Panel on Climate Change (IPCC). The IPCC is best known for the publication of its assessment reports, the first of which appeared in 1990. Since then, three other reports have been issued—in 1995, 2001, and 2007—and all of them have provoked criticism and occasionally acrimonious debate. Why? Because of the reports' insistence that human activity is contributing significantly to an increase in the concentration of greenhouse gases in the earth's atmosphere. This conclusion was reinforced by the second of the assessment reports, and that led to a broadside by a past president of the National Academy of Sciences (and health consultant for R.J.Reynolds Tobacco Company) who claimed that he had "never witnessed a more disturbing corruption of the peer-review process than the events that led to this IPCC report." [5]

---

[5]This led, in turn, to a defense of the IPCC by the presidents of the American Meteorological Society and University Corporation for Atmospheric Research who deplored "systematic effort by some individuals to undermine and discredit the scientific process . . . [which has concluded] that there is a very real possibility that humans are modifying Earth's climate on a global scale. Rather than carrying out a legitimate sci-

The third assessment report, commonly known by its acronym TAR, hammered in the conclusions of the two previous reports. The planet was indeed warming and human activities were leading to a probable change in climate. Criticism came this time from an insider, an MIT professor who had himself contributed to a part of the report. Professor Richard Lindzen complained that there were "numerous problems with model treatments." And again, a rebuttal, this time from Sir John Houghton, chair of the group in which Lindzen had worked, who noted that Lindzen had previously expressed satisfaction with TAR.

The fourth assessment report (AR4) continued the jeremiad. What did it say? Principally that:

- Earth's climate is warming unequivocally;
- More than 90 percent of the increase in greenhouse gas concentration is due to anthropogenic (man-made) sources;
- This will continue for centuries even if greenhouse gas concentrations were to stabilize.

The IPCC itself, and its findings, have been blasted in the court of media coverage. Why? One possible answer is what one commentator sees as the indifference of the American consumer who "each year seems to know less, and care less, about how much energy he or she uses, where it comes from, or what its true costs are."[6]

But—really—who reads the IPCC's assessment reports? These publications are designed for policy makers, not the general public. The substance of the Panel's work filters through to the general public through popular literature. Issues like peak oil, carbon footprints, global warming, water and material shortages have all spawned classics in the past 50 years. Simply to cite the major papers and books written on these subjects since 1960 would fill this chapter. When did environmental awareness burst onto the media stage of the US? It did not 'burst' so much as it came gradually, creeping in on cats' paws, as there dawned on people a realization of the consequences which inevitably followed our abuse of the land, the water, and the air. Not that there weren't some wake-up calls. First of these might be Bill McKibben's *The End of Nature* in which he wrote that "a ninety-centimeter rise in sea level sounds less ominous than a one-yard rise; and neither of them sounds so ominous until one stops to think that over a beach with a normal slope such a rise would bring the ocean ninety meters (that's 295 feet) above its current tide-line."[7]

---

entific debate. . . .they are waging in the public media a vocal campaign against scientific results with which they disagree."

[6] Paul Roberts, *The End of Oil*, Houghton Mifflin Co., 2004.
[7] Bill McKibben, *End of Nature, The*, New York: Random House, 1989.

When McKibben's book first appeared, the atmosphere contained 360 parts per million carbon dioxide. Twenty years later, it had risen to 390 parts per million. The world uses more and more energy, with the largest increases possibly coming from the use of coal. "That is bad news," McKibben writes, "since coal spews more carbon dioxide into the atmosphere than any other sort of energy."

There will be a need to revisit the issue of coal-generated power later in this book. But to follow the thread of McKibben's exposition, one needs to list the next source of atmospheric pollution that he introduces: methane—which "when it escapes into the atmosphere is 20 times more efficient than carbon dioxide at trapping solar radiation and warming the planet."

Prediction of global warming and other environmental phenomena is done with computer simulation. McKibben was writing in the late 1980s since which time mathematical modelling and its computer implementation have become more sophisticated. The models available to McKibben in 1989 predicted that "when the level of carbon dioxide or its equivalent in other greenhouse gases doubles from pre-Industrial Revolution concentrations, the global average temperature will increase, and the increase will be 1.5 to 4.5 degrees Celsius, or 3 to 8 degrees Fahrenheit."

Early in 2010, McKibben published an article about the Copenhagen conference on global warming.[8] In his piece, McKibben wrote:

> Rajendra Pachauri, the UN's chief climate scientist, endorsed the goal of reducing carbon levels to 350 parts per million, even though it went far beyond what his IPCC colleagues had concluded just two years earlier. 'What is happening, and what is likely to happen, convinces me that the world must be really ambitious and very determined at moving toward a 350 target,' he said.[9]

By the final days of that conference, 112 countries had endorsed the 350 target, but, as McKibben noted, "they were the wrong countries—they included mostly the small, poor countries who have little economic or political clout." President Obama's appearance at the conference raised hopes that there would be some dramatic turnabout in the stalemated talks, but from the beginning of his address, it became clear that there would be no breakthrough. "Instead of the generous and open Obama of the campaign (McKibben wrote), there was a pinched and stern Obama, lecturing the assembled heads of state." What the president managed to

---

[8] Bill McKibben, "Heavy Weather in Copenhagen," *NY Rev of Books*, March 11, 2010.

[9] In 2007, the year of the Greensburg tornado, McKibben founded a group called '350.org' which advocated a return to carbon dioxide concentrations in the atmosphere approximating the levels which have prevailed over the last few thousand years.

wrangle from the heavy hitters at the conference—China, India, South Africa, and Brazil—was called 'the Copenhagen Accord.' Three pages long and vague about target dates, it appeared to commit China to something even if that something wasn't spelled out. Obama had to admit that the conference had been a dud: "I think that people are justified in being disappointed about the outcome." According to McKibben, "the verdict was that the conference had failed spectacularly." He went on to say:

> [The] clash between political realism and scientific realism will be at the center of climate policy for many years to come. Campaigners had hoped that science might trump politics at Copenhagen, and instead the opposite happened. Obama, and the Chinese leadership, won't try to force the pace for their political systems, at least in dramatic ways.

In 1972, there appeared the first of three books entitled *Limits to Growth* (LTG). This was followed by a 20-year update, *Beyond the Limits* (BTL, 1992) and finally *Limits to Growth—The 30-year Update* in 2002.[10] The original book was produced at MIT by four colleagues—Donella Meadows, Jorgen Randers, Dennis Meadows and William Behrens III—who had been commissioned to analyze aspects of growth in world population and economy by an informal group called the Club of Rome. A major component of the scientists' original work was a computer model called World3. An updated version of World3 was used for the 20-year update.

As its title suggests, LTG argued that ecological constraints would have a profound effect on global development. The end of growth appeared, in 1972, to be a distant phenomenon. All the scenarios developed by World3 indicated growth well into the next century. Growth, assessed in 1972, appeared to be "comfortably below" the carrying capacity of the planet. By 1992, the situation had changed. By that year, the world's population had overshot the Earth's capacity to support it. That is why the authors decided to give the 20-year update its title—Beyond the Limits. Overshoot, as the authors called this imbalance between people's demand on their environment and what the environment could sustain, was now a fact. The issue had become how to bring humanity's environmental demands back to sustainable levels.

The publication date of *BTL* coincided with that of the Rio de Janeiro conference on environment and development. The goals of that conference were not met, and the results of its follow-up conference, Rio+10, were even more disappointing. Said the authors of *LTG—The 30-Year Update*: "[The conference] was almost paralyzed by a variety of ideological and economic

---

[10]Donella Meadows *et al.*,"Limits to Growth: The 30-Year Update," Chelsea Green Publishing Company, 2004.

disputes, by the efforts of those pursuing their narrow national, corporate, or individual self-interest."

Donella Meadows died before the 30-year update appeared, but hers is the last chapter: *Tools for the Transition to Sustainability*. The term 'sustainability,' one might add, was coined by the World Commission on Environment and Development (1987), commonly referred to as the Brundtland Commission. Gro Harlem Brundtland, a former Prime Minister of Norway, was its leader.

The 30-year update of *LTG* includes a graph depicting the *ecological footprint* of humanity versus the unchanging carrying capacity of the planet. Ominously, the curve representing the ecological footprint overtakes the (horizontal) carrying capacity line in the 1980s. By 2000, humanity's demand exceeded the Earth's supply by 20 percent, *i.e.*, by the dawn of the millennium it took 1.2 planets the size of Earth to satisfy people's demands on the environment.[11] The authors of *LTG—the 30-year Update* were much more pessimistic than they had been in 1972. "It is a sad fact that humanity has largely squandered the past 30 years in futile debates and well-intentioned but halfhearted responses to the global ecological challenge," they wrote. "Much will have to change if the ongoing overshoot is not to be followed by collapse during the twenty-first century."

Limiting growth was not a popular concept with politicians and businessmen. But as time passes, it has become clear that the constraints imposed by finite resources are not to be ignored. What have been called the "highly aggressive scenarios of World3" have proven to be astonishingly accurate.

In the 2002 update, the authors of *LTG* admit that, in their earlier books, they failed to attain all their goals. One of these was to get people to question "the pursuit of growth as a panacea for most problems." They did introduce the term "limits to growth," but they feel that this term has been misunderstood and employed, typically, in a simplistic way. They anticipated a collapse in global resource use. "The collapse will arrive very suddenly, much to everyone's surprise," they wrote. "And once it has lasted for some years, it will become increasingly obvious that the situation before the collapse was totally unsustainable."

In the mainstream of environmental writing, the situation is clearly viewed as becoming more dire. This comes through in the writing of Lester Brown who has written a series of books with titles which include the words *Plan B*. The one which appeared in 2008, *Plan B 3.0*, has for its subtitle *Mobilizing to Save Civilization*. This speaks volumes about the author's

---

[11] The scientists who developed the notion of 'ecological footprint' defined it to be the land area required to support people's needs and to absorb their carbon dioxide emissions.

assessment of how things are going.[12] The subtitle for *Plan B 2.0*, which appeared two years before, had been "Rescuing a Planet Under Stress and a Civilization in Trouble." Brown is a MacArthur Fellow. In his preface, he includes an anecdote about Elizabeth Kolbert, who once interviewed Amory Lovins, another MacArthur Fellow and, according to *Time*, one of the 100 most influential people in the world (2009). When Kolbert asked Lovins about thinking outside the box, Lovins replied, there is no box. "That," Brown wrote, "is the spirit embodied in Plan B."

The deterioration in the Earth's health was accelerating. In the list which *Foreign Policy* labels "Failed States," the number of countries with scores exceeding 100 (120 represents total failure) had increased from 7 in 2005 to 12 in 2006. There was an avalanche of bad news: Peak oil "could be on our doorstep." By 2030, China's cars might be consuming more oil than the world production in 2008. Could the US phase out its coal-fired power plants before the melting of the Greenland ice cover became irreversible? Ten years before the book's appearance, the Yangtze River valley in China had experienced flood damage greater in value than the country's annual rice crop. (As a result of the flooding, China banned logging in the Yangtze River basin.)

Brown cites examples of previous civilizations—Sumerian and Mayan—destroyed by environmental mismanagement. "A team of scientists led by Mathis Wackernagel concluded in a 2002 study . . .that humanity's collective demands first surpassed the earth's regenerative capacity around 1980," he writes.[13] "Today, global demands on natural systems exceed their sustainable yield capacity by an estimated 25 percent."

Fifteen of twenty-four "primary ecosystem services"—including oceanic fisheries—were being degraded or headed toward collapse. Ditto, tropical rainforests. It had been clear for some time that carbon emissions needed to be cut, but not a single country on earth had achieved carbon neutrality. World grain production was lagging behind consumption, with the result that "corn prices nearly doubled and wheat prices tripled between late 2005 and late 2007."

"Just when it seemed that things could not get much worse, the United States, the world's breadbasket, is planning to double the share of its grain harvest going to fuel ethanol—from 16 percent of the 2006 crop to 30 percent or so of the 2008 crop," writes Brown. "This ill-conceived US effort to reduce its oil insecurity has helped drive world grain prices to all-time highs, creating unprecedented world food insecurity."

To save civilization, Brown argues that Plan B has to be implemented with a speed comparable to that which brought about the restructuring

---

[12] Brown's book was published by W.W.Norton & Company in 2008.
[13] This was the study which *Limits to Growth–30 Year Update* alluded to in comparing our ecological footprint with the carrying capacity of the environment.

of the US economy in 1942, when the country entered World War II. The components for that plan are already in place: Wind turbines to generate electricity; reforestation projects like one in South Korea which resulted in a tree cover for 2/3 of that country, stripped bare during the Japanese occupation (trees help to sequester carbon dioxide in the atmosphere); fish farming techniques developed in China; soil erosion control (as practiced in the US); reduction in the use of cars . . . The list goes on.

The last of the road signs in this romp through the recent environmental literature is Paul Roberts' *The End of Oil*. It first appeared in 2004 and was reissued a year later.[14] In a nutshell, the problem that Roberts deals with is our appetite for hydrocarbon-based energy and the way it continues to outstrip the world's supply of easily accessible oil. Our energy economy has hit a peak, he writes, and with every passing year, the "extraordinary machine we have built" to slake our thirst for oil cannot sustain itself, at least in its present form. " Not a day goes by without some new disclosure, some new bit of headline evidence that our brilliant energy success comes at great cost—air pollution and toxic waste sites, blackouts and price spikes, fraud and corruption, and even war."

Roberts talks of this country's "SUV-driven demand" for oil. To satisfy that demand, the US must import oil from foreign sources, many of whom regard the US as their enemy. Oil is not the only problem. Twenty-six percent of our energy comes from coal, a four-letter word for environmentalists because it is "fatally dirty." In Roberts' opinion, our energy economy is deeply flawed in nearly every respect. The oil industry, he points out, "is among the least stable of all business sectors, tremendously vulnerable to destructive price swings and utterly dependent on corrupt, despotic 'petrostates' with uncertain futures." Even worse, it is now obvious to all "but a handful of ideologues and ignoramuses" that our dependence on fossil fuels is related to climate change, specifically the rise in global temperatures. "If left unchecked," he writes, "this so-called greenhouse effect will keep warming the earth until polar icecaps melt, oceans rise, and life as we know it becomes impossible."

From 80 million barrels a day, demand for oil will jump to 140 million barrels a day by 2035. Where is the additional oil going to come from? Roberts points out that some reservoirs, yet untapped, like the Arctic, will be expensive to develop. Ominously, demand for electricity in the Third World has risen so sharply that countries like India and China have sidelined environmental concerns to build hundreds of coal-fired power plants "whose emissions may make it impossible even to slow climate change." And these are not the worst cases!

---

[14]Excerpts from *The End of Oil: On the Edge of a Perilous New World* by Paul Roberts. Copyright (c) 2004 by Paul Roberts. Reprinted by permission of Houghton Mifflin Harcourt Publishing Company. All rights reserved.

It is clear that a transition from our current energy technology to a new one is going to occur. The question is whether this will be an orderly or a chaotic transition. How long will it take? What will it cost? And how does it all start? These are Roberts' questions. And this is the time to ask these questions, when the US and Great Britain "are struggling to extricate themselves from a *second* oil war in Iraq that, whether openly acknowledged or not, was clearly meant to restore Middle Eastern stability and maintain Western access to a steady supply of oil." American efforts to ensure access to the precious fuel have engendered "political instability, ethnic conflict, and virulent nationalism in that oil-rich region." Roberts notes that, " even before American tanks rolled into Baghdad to secure the Iraqi Ministry of Petroleum, leaving the rest of the ancient city to burn, anti-Western resentment in the Middle East had become so intense that it was hard not to see a connection between the incessant drive for oil and the violence that has shattered Jerusalem, the West Bank, Riyadh, Jakarta, and even New York and Washington."

Roberts has some scathing remarks to make about consumers who "think nothing of buying ever larger houses, more powerful cars, more toys and appliances—increasing their energy use without even knowing it."[15] But signs of change are everywhere as "an exhausted system" gives way to something new: Oil companies are reinventing themselves to sell natural gas. Skirmishes are being fought over climate policy. And countries are racing each other to access the "last big oil."

How much of the stuff is left? Oil depletion, in Roberts' estimation, "is arguably the most serious crisis ever to face industrial society." Proven reserves stand at around 1.7 trillion barrels, and about half of that is in the Middle East. Oil optimists believe that approximately 1.5 trillion barrels remain to be discovered. But where? There remain few places on earth, Brown remarks, where oil could be found but where oil companies have not already explored. Undiscovered reservoirs are more difficult to access, and they are likely to be smaller than the mega-fields of Saudi Arabia and Russia. Since 1995, the world has consumed 24 billion barrels of oil a year but has found, on average, 9.6 billion barrels of new oil annually. And oil pessimists, citing what they feel to be more realistic forecasts of oil discovery, believe that the oil remaining to be discovered is of the order of 1 trillion barrels (not 1.5 trillion as the optimists hold) and that the peak occured . . . in 2010.

****

Any nay sayers? There are the politicians like Oklahoma's Senator James Inhofe who has decried climate change as "the greatest hoax ever

---

[15] The quote earlier in this chapter about the obtuseness of the US general public comes from Roberts.

perpetrated on the American people."[16] The senator's statement has been castigated by scientists like John Terborgh. In his review of a book by Tim Flannery, Terborgh writes that such an opinion "would be farcical if attitudes such as this weren't driving our public policy on an issue that threatens the well-being of billions of people."[17]

There are, of course, scientists who pooh-pooh the significance of global warming. The most celebrated of these is probably British-born Freeman Dyson whose contrarian views on climate change have landed him on the cover of the *New York Times Magazine*. He read *The Wonderful Wizard of Oz* as a boy, got a sense of the US as an "exciting place where all sorts of weird things could happen," and emigrated to the US after World War II.

Dyson is an intellectual giant in the world of theoretical physics. But his pollyanna-ish view of carbon dioxide (it helps plants to grow) cost him friends. He began to speak out in 2005 on what he perceived to be unwarranted worry about climate change. "All the fuss about global warming is grossly exaggerated," he claimed. Climate change had become an "obsession" of the practitioners of a secular religion called environmentalism. "It's always possible Hansen could turn out to be right," Dyson said of climate scientist John Hansen of Columbia University and NASA. He went on to add:

> If what he says were obviously wrong, he wouldn't have achieved what he has. But Hansen has turned his science into ideology. He's a very persuasive fellow and has the air of knowing everything. . . . By the public standard he's qualified to talk and I'm not. But I do because I think I'm right. [18]

Hansen was contemptuous of Dyson's remarks. "There are bigger fish to fry than Freeman Dyson," he told the *New York Times*. In an e-mail he subsequently sent, he wrote that Dyson should first do his homework, which he clearly hadn't on global warming, if he were "going to wander into something with major consequences for humanity and other life on the planet."[19]

---

[16] Or is it the second greatest, after separation of church and state? Surfing the web and transiting through at least one hostile web site left room for doubt. In any case, the Senior Senator from Oklahoma has no patience for global warming. See his floor speech, "Climate Change Update," Jan. 4, 2005, available on the web at *inhofe.senate.gov/pressreleases*. His adamancy on climate change won him a slot on *Rolling Stone*'s list of 17 "Climate Killers" as *God's Denier*. See the magazine's issue 1096, Jan. 21, 2010.

[17] John Terborgh, "Can Our Species Escape Destruction?", *New York Review of Books*, Oct. 13, 2011.

[18] See Nicholas Dawidoff's article about Dyson, "The Civil Heretic," which appeared in the *New York Times Magazine*, March 29, 2009.

[19] *Ibid.*

Dyson is an octogenarian maverick. He loves to take his scalpel to a congealing consensus. Al Gore's *An Inconvenient Truth* left him unimpressed. Of the former Vice President, Dyson said, "He certainly is a good preacher."

****

This is the story of Greensburg: how the town was destroyed; how it chose to come back. There are heroes in this story—although those wearing the white hats sometimes lose them in the wind—and a few hissable villains. Mostly these are ordinary people trying to cope under horrendous circumstances. Greensburg, arguably a microcosm of the nation, offers a model—a flawed and self-questioning one, at that—of how to survive in difficult times. Times which will grow ever more difficult as the planet's energy and environmental problems swell from molehills to mountains. Can we learn from what has happened here? From the town's achievements as well as its failures? Those are the questions. Our answers will determine our future.

A LITTLE HISTORY

"Where is Kansas?" asked the man, in surprise.

"I don't know," replied Dorothy, sorrowfully; "but it is my home, and I'm sure it's somewhere."

–Wonderful Wizard of Oz

Greensburg, a town described at the end of the nineteenth century by a local paper as "the liveliest town in the state today, for money, marbles or watermelons,"[20] lies astride US Highway 54. It is located about 50 miles southeast of Dodge City and 30 miles west of Pratt. Both are much larger than Greensburg, the county seat of Kiowa County. All three towns—Greensburg, Dodge, and Pratt—lie in the Great Plains of America, a region originally covered by native grasses, neither forest nor desert, but something in between, a savannah buffering the lush valleys of the Mississippi and its tributaries to the east and the Rockies to the west. The region is frequently buffeted by fierce volleys of wind. Kansas is, in fact, the third windiest state in the nation.

"It's so flat," Matt Deighton, one of the town residents who figures in our story, once said. "If you squint your eyes and look east, you can see two days coming." He made this remark to a BBC reporter. *Why* a reporter from the BBC should be visiting a small town in south-central Kansas is the subject of this book.

Greensburg was named after a stagecoach driver, D. R. "Cannonball" Green. Green's coaches were called *The Cannonball Stageline*, and that's how Green got his nickname. A sign just to the west of Greensburg informs motorists that this stretch of US 54 is known as The Cannonball Stageline Highway. Father Time himself, Green boasted, could not keep up with his

---

[20] *Kiowa County Signal*, June 1, 1888

stagecoach. Surveyors laying out the town site in 1884 persuaded Green to make the new town a stage stop. In return for this, he was given free land and the promise to name the new village after him.

Cannonball Green had little to do with the development of the town named after him. He became fairly wealthy and bought a farm. There he raised championship horses. The enterprise came to grief, however, when Green's son, in an attempt to control rodents, doused a pack of rats with kerosene and set them afire. Before dying, one of the rats ran into the barn and set it ablaze, destroying the building and all the horses inside their stalls.

There is a story which claims that Cannonball Green evicted Carrie Nation from his coach after she swatted a cigar from his mouth.[21] This tale probably evolved from an incident which was reported on February 6, 1884, in the *Kinsley Graphic*. The paper's article begins with a quotation from another newspaper, and it reads as follows:

> A slight unpleasantness occurred on The Cannon Ball last Thursday on the route from Kingman to Pratt. Capt. H. H. Patten, of Greensburg, Dr. Dickerson, of Saratoga, and Mrs. Hinkle, of Kingman, were passengers. The learned knight of the powder and the pill . . . persisted in smoking and that too of a weed, the fume of which was vile enough *to strangle a glandered horse*. Mrs. Hinkle complaining that the smoke rendered her quite uncomfortable, Capt. Patten requested the Doctor to desist, but to this he paid no sort of heed. A few—and to the credit of the Captain be it said, but very few—words of uncomplimentary import passed, when the Captain reached out his good right arm, and . . . knocked the obnoxious weed out of the doc's mouth.
>
> —*Pratt County Press*

This passage gives a sense of the literary arabesques and editorializing which jazzed up Great Plains journalism of the day. It also reminds us of the social distance we have come, from a time in which smoking in public was tolerated—at least up to a point—to current restrictions.

A glandered horse? Glanders is a disease, one of whose symptoms is excessive discharge of mucous from the nostrils.

---

[21] Internet apocrypha. Ms. Nation did not go on her rampage against saloons until the following century.

H.H. Patten—the "H.H." stand for "Harrison Henry"—was secretary and treasurer of the Greensburg Town Company, which owned the town site. Like William Hinkle, the president of the company and the husband of the woman in the stagecoach who had been nauseated by cigar smoke, he was a Civil War veteran. Patten had actually captained a contingent of African-American troops in the Union Army. Patten, a lawyer, and Hinkle, a businessman, were movers and shakers of early Greensburg.

One of the town's newspapers, the *Kiowa County Signal* (*KCS*) appeared for the first time in 1886. The newspaper has continued, although not always under exactly that name, to the present. Since the Historical Museum of Kiowa County was flattened by the tornado, and its collections unavailable, the *Signal* is a major source of information for this book. The Greensburg office of the newspaper suffered damage during the tornado of May 4, 2007. Water-damaged back copies of the weekly were stored temporarily in the basement of a sister publication, *The Pratt Tribune*. Fortunately all back issues were also preserved on microfilm at the Kansas State Historical Society just outside Topeka, the state capital.

A number of newspapers have served Kiowa County over the years. One such paper was the *Greensburg Republican* which amalgamated with another Greensburg paper, the *Consolidated*, in January, 1913. The masthead of the joint publication appeared on this weekly for the next six years. In the 1920s, both papers merged with the *Signal*. Tracking the mergers and closings of newspapers in south-central Kansas might be a subject for a Master's thesis.[22]

In its first year of publication, the *Signal* ran aphorisms from Antiquity (Marcus Antoninus, Seneca) and from more recent times (Alexander Pope, William Penn, Benjamin Franklin). In its edition of February 26, 1886, these ran beside advertisements for Brown's Iron Bitters ("Strengthens the Muscles, Steadies the Nerves, Enriches the Blood"), Allen's Lung Balsam ("25 cents for Croup"), and Prickly Ash Bitters ("Cures All Diseases of the Liver, Kidney, Stomach and Bowels").

Ed Schoenberger was elected curator of the Kiowa County Historical Society in 1993, five years after he became sexton of the Greensburg cemetery. A wiry and energetic sexagenarian, Ed agreed to be interviewed on a rainy afternoon in June, 2008. His dimly lit office stands on the edge of the cemetery, overlooking neat rows of what were that day glistening-wet tombstones. Schoenberger was wearing a Pro-Life T-shirt. A George W. Bush sticker was posted on the wall of his office.

---

[22]University of Kansas historian, Bill Tuttle, made available one such thesis. Its newspaper references were particularly helpful.

"Once something gets into print, it's difficult to change it," Schoenberger observed, citing in particular the story of Cannonball Green and Carrie Nation. "I've had people get angry at me because my research contradicts what they have read about a subject."

****

Names in Kiowa County reveal the ethnic origin of the region's settlers. These do no differ much from other parts of the Midwest. German-Scots-Irish-English names predominate.[23] But one group of settlers should be singled out. The Mennonites settled in Kiowa County at the beginning of the twentieth century. The simple life style of the stricter Mennonites is consonant with green philosophy, and the sect's relief efforts played a crucial role in the aftermath of the tornado that levelled Greensburg.

The Mennonites are a group with Anabaptist roots in northern Germany and the Netherlands. Named after one of their pastors, Menno Simons, they adhere to a doctrine of non-resistance which entails a refusal to take up arms, to become involved in litigation, or even to hold public office.

Persecution has made of the Mennonites a migratory flock. In 1763, Catherine the Great of Russia issued a proclamation, inviting farmers from Western Europe to settle in the south of Russia on land won in a war with Turkey. On March 3, 1788, Mennonite emissaries obtained a charter of privileges from Catherine, including "complete religious freedom and exemption from military service for all time."[24] Soon afterwards, groups of Mennonites began leaving Western Europe to settle in the south of Russia.

The Mennonites never assimilated into Russian culture. Besides retaining their own religion, they continued speaking low German, or *Platt Deutsch*, among themselves. Pressure from the Russian government to integrate and become full-fledged Russian citizens began to mount. A law of 1874 which required universal military service of all Russian subjects heightened the Mennonites' fear of assimilation. They dispatched several delegations to St. Petersburg to request a continuation of their exemption. They did manage to wrest from the government an offer of alternative service. Yet even this was insufficient to allay the Mennonites' fears. About 18,000 of them decided to move on once again, this time to North America. One group of Mennonites settled first in Ontario. In 1893, some of the sect relocated in the Canadian provinces of Manitoba and Alberta. After the Bolshevik revolution and the imposition of Communist rule in Russia,

---

[23] *www.city-data.com* lists the ancestral make-up of Greensburg as 31.4 percent German, 11.8 percent English, and 11.1 percent Irish.

[24] Cornelius J. Dyck, *An Introduction to Mennonite History*, 3rd ed. (Scottdale, PA: Herald Press, 1993).

heightened repression drove more Mennonites out of the Soviet Union to join the existing Mennonite communities in the US and Canada.

The first Mennonites to move to the region just south of Greensburg arrived in 1908. These included, eventually, descendants of a minister, Tobias Unruh, one of the members of the so-called "deputation of twelve" who had come to the US in 1873 in search of land where they could settle. The family name 'Unruh' is common throughout Kansas. A cell phone directory for Greensburg, appearing in July, 2008, has—apart from three government agencies—nothing but Unruhs listed under the Us.

There is a photograph of Tobias and his immediate family in *The Tobias A. Unruh Biography, Diary and Family Record* which was still in the possession of his great great grandson, Gordon Unruh. The photo shows him to be stiff and glassy-eyed, but this is not uncommon in nineteenth century photography. The *Biography* says that, "he was an intensely spiritual man, one whose life was lived in complete submission to God."

Unruh kept a journal of his voyage to America in which he wrote on May 22, 1873 that:

> Great waves were thrown against the ship. Water was a foot deep on deck. I sighed with Peter: 'Lord Save Us.' True, the comforting hand of God was watching over us. 'Fear not for I am with thee, in the midst of the stormy sea.'

An interview with President U.S. Grant in which Tobias asked for a 50-year exemption from military service for Mennonite men was inconclusive. But he still returned to Europe under the impression that the Mennonites would be exempted from military service *in perpetuity*. What exactly transpired between the president and the Mennonites? In Tobias' journal, the relevant entry reads as follows:

> August 8. We left New York 8 a.m. to see the president. At 8 p.m., we were introduced to the president in his residence. Our agent Hiller sent his servants to him with a message and also our petition and plea. He looked them thru, then requested that we come and meet him. He was a plain man and very friendly. He informed us that the constitution has a concession that it will not over-ride a man's conscience and religious freedom is guaranteed. We appreciated this information, expressed our gratitude and bid him adieu.

Land was available to the Mennonites as a result of the Homestead Act of 1862 and the development of the railroads, which made virgin land in states like Kansas accessible to new settlers.

Tobias himself settled in South Dakota in 1875, and there he died, shortly afterwards, of typhoid. His son, Benjamin, died of the same cause within months, but Tobias' grandson, John B. Unruh, lived till 1943, in which year he died in McPherson, Kansas. In 1878, John Holdeman, a Mennonite reformer from Ohio, who held that the church had become corrupted, baptized 78 Mennonites in McPherson. Three years later he baptized an even larger number of Russian emigrants in Manitoba. The Holdemans, as the offshoots of the Old Mennonite Church are sometimes called, are characterized by the strictness of their faith. Their web page[25] contains this passage which prescribes how members of the group are to conduct themselves:

> The families [of the sect] live a plain and simple lifestyle. They do not have television sets, radios, or CD players. They have telephones and many of them use the fax machine for communication. They drive plain, single color cars and mini-vans that are not sporty and high class, being careful to remove the radio systems of all vehicles. The dress is of similar uniform design with the allowance of dresses to have individual patterns and colors of fabric. The women wear an everyday head covering of a small black cap that is placed over a bun of hair set on the back of the head.

The 'simple lifestyle' mentioned in the first line of the quote shuns excesses, even unnecessary personal possessions. Mennonite communities, *e.g.*, those in Russia, were admired for their self-sufficiency. The way Mennonites live exemplifies the sustainability advocated by the green movement as it unfolded in post-tornado Greensburg.

Tobias' great-grandson, Isaac, went on to Greensburg, in 1930. The Mennonite community south of town, known as the Bethel Congregation, was then growing in number and thriving. Four years before Isaac's arrival in their midst, the Holdeman Mennonites there had already erected their own church.

The Mennonites have been active in relief operations for more than half a century. In some ways, their work immediately after the Greensburg tornado, under the auspices of either Christian Disaster Relief (Holdeman Mennonite) or Mennonite Disaster Services (an offshoot of the Holdemans), was among the most effective efforts of the relief organizations to come to the stricken town's aid. In the course of his interview for this book, Lloyd Goossen, a descendant of Goossens who had settled first in western Canada, said that he was preparing to drive to Iowa, suffering in the spring of 2008 from flooding. He would be accompanied by several of his fellow Holdeman

---

[25] *www.holdeman.org*

Mennonites from Greensburg. This was only a year after the tornado had totally destroyed his own home and workshop. The Mennonites were going in order to help Iowans contend with a natural disaster that had inundated towns along the shores of swollen rivers. At the time of his departure for Iowa, the foundation of Lloyd and Anita Goossens' new house had just been laid in the prairie, beside Goossen's reconstructed workshop.

****

For a town of its size, Greensburg had more than its share of tourist attractions. Two sites and one item of interest were mentioned in pre-tornado tourist brochures.

The first of these was the world's largest hand-dug well. Begun on August 9, 1887, the well provided water for the city and its industries. Only later was well water used to supply the steam engines of the trains passing through town, although this is often cited as the original purpose of the well. Confusion may have arisen because two of the directors of the Greensburg Water Supply and Hydraulic Power Company which dug the well worked for the Atchison, Topeka, and Santa Fe Railway. Measuring 109 feet deep and 32 feet in diameter, the well provided Greensburg with water until 1900.

In 1916, the well was cleaned up and a water tower erected. By the following year, the well was again supplying the city with water. But in 1932, the State of Kansas banned open wells, and the Big Well was closed down. Seven years later, it was opened to tourists who were allowed to climb down to the bottom of the well. By then, the original helical staircase, supported by rail irons driven into the well wall, had been replaced by a zigzag staircase. A small museum adjacent to the well contained the second of the town's tourist attractions, a 1000-pound meteorite which was uncovered by a local resident, H. O. Stockwell, in 1949.

The third and last of the town's pre-tornado attractions was the Twilight Theatre. This building owes its origins to a fire which the *Greensburg Republican Consolidated*, in its edition of January 9, 1913, covered under the headline, "East Side of Main Street Swept by Fire." Five business buildings, one of which was the original Twilight, were destroyed. Loss on buildings and stocks was reckoned at about $15,000. This was only partially covered by insurance.

Rebuilt of brick three years later, the Twilight was first called the Miller-Wacker Auditorium. A man by the name of Charles Spainhour rented the Auditorium in 1917. About the same time, he bought out a rival theatre, The Empress. This second theatre was located further south on Main Street and had been unaffected by the 1913 fire. Spainhour bought the theatre to close it down. He wanted to stifle competition with the Auditorium. On May 3, 1917, the *Signal* offered its opinion that "now that we have a fine building in which to furnish entertainment for the public, it is the duty of the public to support it loyally."

A 1918 photograph of the theatre shows a two-story building with a porch and balcony which runs the length of the structure on its Main Street elevation. A photo taken more than half a century later shows little change in the exterior appearance of the theatre. Spainhour showed silent films in the Auditorium, accompanied by music from an Edison music box. His son, Ben, sold peanuts and popcorn to the audience. In 1919, Spainhour bought the building. At that time, there was only one movie projector in the Auditorium. After one reel of a movie was shown, the projector was rewound and the next reel mounted while the audience watched a slide show of advertisements.

During the winter months of 1918-1919, the theatre was shut down because of the influenza epidemic sweeping the country. Within years of its reopening, Spainhour began showing movies every day of the week, with matinees on Saturday and Sunday afternoon. In 1923, the theatre was completely renovated, new projection equipment installed, and a name-the-theatre contest held. The result of that contest was a name change for the Miller-Wacker Auditorium which, henceforth, would be known as the Twilight Theatre. Again from the *Signal*, improvements led to "absolutely flickerless, projecting pictures free from eye strain, a feature seldom found in any but the largest theatres." The new theatre boasted an embossed tin ceiling. The fragments that survived the tornado were stored in a workshop adjacent to Ed Schoenberger's office at the Greensburg cemetery.

Five years after the new theatre's 25th birthday, in 1947, Charles Spainhour died. Management of the theatre was bequeathed to his son, Ben. Ben's son, Con, took over in 1971. Toward the end of the 1980s the Twilight began to falter and finally shut down entirely. This happened shortly before Farrell Allison, an agricultural consultant from the Texas panhandle, moved to Greensburg with his wife, Debbie. The couple was instrumental in setting up a 501(c) non-profit corporation to buy and maintain the theatre. Farrell became the president of the theatre board. The building which that board took over could accommodate 400 to 425 people at the time. Early in the twentieth century, it had been the biggest theatre west of the Mississippi. But the Twilight—its big screen and stage notwithstanding—was obsolete. Each of its bathrooms could hold only one person at a time. Both of them lacked sinks. The sound system was out of date. So were the building's heating and cooling systems. To renovate the theatre, the board concluded in late 2006 that it would need to raise $357,000. The state of Kansas authorized the board to sell $250,000 in 70 percent tax credits. This meant that anyone contributing $1,000 to the theatre project would receive

$700 in tax credits. By year's end, the board had sold about $15,000 in tax credits, a far cry from what needed to be raised by the end of 2007.[26] The tornado which struck Greensburg that year put the quietus to this project since the building was wiped out on May 4, 2007.

****

Like other towns in the Great Plains, Greensburg slowly shrank from 1960 onward. At its apogee that year, the town's population was 1988. By 2006, that population had shrunk to 1389, a loss of more than 30 percent.[27] Young people were moving out in search of greater economic opportunity elsewhere, leaving behind an aging population. Their demands on public services aggravated the problems faced by the town.[28] By most accounts, Greensburg was a dying community when it was destroyed in the spring of 2007. The *New York Times* described the tornado as Nature's *coup de grace* on a terminally ill town.

Many Greensburgers were acutely aware of the situation and were prepared to do something about it. All of these people were to play important roles in the recovery of the town after May 4, 2007.

First up: Lonnie McCollum, a state trooper who had spent his youth in Greensburg. He and his wife, Terri, returned to the town after he retired from the force in 2004. Lonnie's career had taken him to Pratt, where he headed up the state police for 12 years, and then to Topeka, where he served as Superintendent of the Highway Patrol. At that time, he became a close friend of Speaker of the Kansas House, Pete McGill. McGill suffered from kidney failure. Lonnie offered to give his friend one of his own kidneys. In 1998, the kidney transfer took place at the Mayo Clinic in Rochester, Minnesota. Lonnie has a photo showing him and McGill in adjacent beds. The operation was credited with helping to prolong McGill's life by five years and to make those five years more comfortable ones. McCollum gave McGill, a Republican with a collection of elephant memorabilia, an elephant's head by the Italian ceramicist, Giuseppe Armani. The gift was returned to Lonnie upon McGill's death. It was one of the few pieces in the McCollums' own collection of Armani ceramics to escape destruction by the tornado.

---

[26] Eric Swanson, "Group raising money to help preserve town's historic theater," *Dodge City Daily Globe*, Dec. 2, 2006.

[27] *www.city-data.com*

[28] *Ibid.* The same web page gives the white, non-Hispanic population of the town as 96.5 percent, the median resident age as ten years more than the median age for the state of Kansas, and the median household income in 2005 as $29,000. The median household income for the state of Kansas as a whole was $42,920.

Lonnie McCollum

McCollum was raised by Depression-era parents who wasted nothing. He admired Native Americans. They shared his parents' prudence and respected the environment. Protecting the environment had long been of paramount importance for McCollum. He ranked it number one on the list of issues facing this country. Although his role in taking Greensburg green was limited to a few weeks, McCollum's commitment to the concept got the ball rolling in those crucial, chaotic days following May 4, 2007.

The McCollums' move back to Greensburg was, in the words of the ex-state trooper, "like taking a huge weight off me." His professional life had been marked by conflict. Retirement to Greensburg meant peace and relaxation. The McCollums built a house on the corner of Florida and Sycamore Streets, close to downtown. They filled it with Armani ceramics and antiques valued by one of their acquaintances at $150,000. "Lonnie's wife," said the friend, "was a compulsive buyer."

Dissatisfaction with the incumbent mayor of Greensburg led some of Lonnie McCollum's friends to persuade him to run for office in 2006. Among them, and possibly the two most persuasive, were Gary Goodman and Mark Anderson. Born and raised in Wichita, Goodman had a long-term friendship with McCollum which went back to the two men's days in Pratt. Pony-tailed and pierce-eared, Goodman does not fit the mold of small-town Midwestern America. He and McCollum, the former state cop, did not always see eye-to-eye, but they were close. Mark Anderson holds a Master's

Ceramic elephant by Giuseppe Armani

Degree in theology and ministered to three Methodist congregations in the vicinity of Pratt. Simultaneously he ran the *Kiowa County Signal* as a one-man band.

    Lonnie McCollum was a reluctant candidate, probably because he could foresee the controversies that being mayor would entail. He had had enough of that as a state trooper. But he caved in to his friends' pressuring in a last-minute and impulsive decision. "Terri was in the shower when I finally persuaded Lonnie to run," said Goodman. "So she was not consulted." McCollum's decision to run came so late that it was only with the assistance of the *Signal*'s editor, Mark Anderson, that the paperwork for his candidacy was filed on time. Lonnie McCollum ran as a write-in candidate. To his astonishment, he won by a 2-to-1 margin over the incumbent, Stan Adolph. Things began to change in Greensburg. "I guess people wanted to go in a different direction," the mayor-elect observed with characteristic modesty.

    One of his first tasks was to appoint a City Administrator. The job went to Steve Hewitt, a son of Greensburg who had attended Pratt Community College, then Fort Hays State University about 90 miles north of Greensburg. An administrator with six years' prior experience by the time he moved back to Greensburg, Hewitt was tall and blonde. His square Midwestern face conveyed transparency and honesty. He was the opposite of Lonnie McCollum in every way, a totally self-possessed Bud Abbott to McCollum's effusive Lou Costello. Hewitt was also a charter member of a group called the Rat Pack which Lonnie McCollum organized upon becoming mayor.

The Rat Pack was a club of sorts—a fraternity sworn to secrecy—which met to hatch schemes to improve the city of Greensburg. Besides McCollum and Hewitt, it included Gary Goodman; Superintendent of Schools, Darin Headrick; Gary Goodheart; and the man who would ultimately succeed McCollum as mayor, Bob Dixson. McCollum's immediate successor as mayor, John Janssen, had been tapped to be brought into the pack, but the tornado overtook that happening. "Dreaming was okay," McCollum said of the Rat Pack. The Packers believed that, somehow, funding for their various ventures would be available, one way or another. It should come as no surprise that all the Rat Pack members went on to play crucial roles in the redevelopment of Greensburg. The four years following McCollum's election were so utterly dominated by Rat Pack ideology that this period in the town's history might be called the Rat Pack years. This is not to say, however, that cracks in the Rat Front did not develop over those years. They did. But the Pack's activism dictated the direction that Greensburg took from 2006 to 2010.

Some of the Rats' projects were pie-in-the-sky notions which the Pack could not pull off: sponsoring Saturday afternoon mime performances on Main Street, for example; or painting a *trompe l'oeil* streetscape on the unsightly rear ends of commercial buildings to improve the town's appearance from the west. But other projects saw the (early) light of day: Starting at 5 a.m., Goodman and McCollum, on their own initiative, painted the facades of 16 derelict or near-derelict buildings fronting US 54.[29]

Still, the demands of holding public office were taking their toll. A week before the tornado, McCollum confessed to Mark Anderson that he was considering resigning. Despite his and Gary Goodman's repainting, for example, there was an on-going problem with the appearance of abandoned buildings in Greensburg. It did not help when some of McCollum's constituents ranted at him in public, accusing him of all manner of personal failings. The tornado acted as a bellows to blow the subterranean smouldering of small-town discontent into a three-alarm fire.

---

[29]Goodman participated despite undergoing quadruple bypass surgery earlier that year.

THE NIGHT IT HAPPENED

> *The house whirled around two or three times and rose slowly through the air . . . . The great pressure of the wind on every side of the house raised it up higher and higher, until it was at the very top of the cyclone; and there it remained and was carried miles and miles away as easily as you could carry a feather.*
>
> *–Wonderful Wizard of Oz*

Spring is the season for tornadoes in Kansas. Warm moist air from the Gulf of Mexico moves north to meet cooler, very dry air flowing west from the Rockies. Numerous tornadoes in the region of south-central Kansas were spotted over a time period from 8:30 p.m. on May 4, 2007, to 2:10 a.m. the following day. Of these, four were "big," and the biggest, most powerful, an EF5 tornado, was the one to strike Greensburg.

The designations EF0 to EF5 refer to the Enhanced Fujita Scale, which measures tornadoes in terms of the strongest winds that they pack. Ted Fujita was a meteorologist who came to the US from his native Japan in the early 1950s to work at the University of Chicago.

Only three months before the Greensburg tornado, the Fujita Scale had been replaced by the so-called Enhanced Fujita Scale. The criteria used to assess a tornado's level of damage were refined. The Greensburg tornado, with wind speeds of 205 mph, was the first-ever tornado with the new EF5 designation to strike a town in this country.

****

The sun shone that Friday, May 4, 2007. Lonnie and Terri McCollum were enjoying the shade of a gigantic maple tree in their yard. The foliage on that tree was so dense that you could stand under it for ten minutes while it rained and not get wet. The McCollums had poured much of their savings into their house, which stood at the northwest corner of Sycamore and Florida Streets.

Alanna Goodman had moved to Greensburg from Las Vegas, Nevada. Six years before, she came with her parents, Erica and Gary Goodman, and her younger brother, Josh. When asked why the family had made such a radical move, Erica quipped that she wanted a more exciting night life. A willowy 22-year-old, Alanna stood out in Greensburg for her quirky fashion sense and her Mediterranean good looks. She lived on Main Street, only a few blocks from the McCollums, in a converted drug store, next door to her vintage clothing store, "Snootie Seconds."

That Friday found Erica Goodman was out of town—like so many other residents of Greensburg, away for one reason or another. There was a forensics meet in Salina, a city 150 miles to the northeast. This siphoned off a number of high school students. Other students were also away that day at athletic meets. Erica Goodman had gone to a Sampler Festival in Garden City where she was promoting Greensburg. Among other items which advertised the municipality was a life-sized cardboard cut-out of Greensburg's soda jerk, Dick Huckriede. Uncle Dickie, as he was known to the townsfolk, had been jerking sodas so long—more than 50 years—that some people actually believed that he had made it into the *Guinness Book of Records*. Alanna had considered working that night on a fashion show she was putting together as a benefit for the Twilight Theatre. Ultimately she decided to go see her father. She left her store in a cross-strapped blouse, a pair of grey shorts and sandals tied at the ankles. The weather was so mild there was no need to dress more warmly.

At 4 p.m., Megan Gardiner, a junior at Greensburg High School, had reported for work at a local restaurant called the Lunch Box. When she took the trash from the restaurant out around 9:15 that night, she noticed "the scariest lightning she had ever seen." Suspecting that her father, a physician assistant (PA) at the Kiowa County Memorial Hospital, or her mother, Julie, would call to tell her to come home because of the impending storm, Megan tucked her cell phone in her bra and washed the dirty dishes as fast as she could.

In November 2006, after living 21 years in Denton, Texas, where he had been employed in the plant department of Texas Women's University, Jerry Diemart returned to his family's hometown of Greensburg. He spent part of the afternoon and early evening of May 4 in Pratt, where he had bought his Mom a mountain bicycle with good shock absorbers. "She's older now and needs those features in a bike, but she still gets around," he said. He and his friend, Andy Huckriede, the nephew of the famous Greensburg soda jerk, left Pratt around 8:50 p.m. with the bike in the back of his Dad's GMC Sierra truck.

State Representative Dennis McKinney lived on the edge of Greensburg. Unlike most of his neighbors, McKinney was a Democrat, the son of Democrats. They had farmed on land between Greensburg and the town of Coldwater, which lies just north of the Oklahoma state line. McKinney had spent much of the day in the fields. Following in his father's footsteps, he farmed when he wasn't legislating in Topeka. But that night he returned home and was watching TV with his 14-year-old daughter, Lindy. Like Gary and Alanna Goodman, they were alone in the house. McKinney's wife was accompanying the high school students who had gone to the forensics meet in Salina. His older daughter was attending the University of Kansas in Lawrence. Like Megan Gardiner (and everyone else in town), they heard the sirens go off.

****

Meteorologist Mike Umscheid was working in his office at the Dodge City airport. He was one of four meteorologists tracking the storms moving that day from Oklahoma into south-central Kansas. Umsheid has the look of a college freshman. In fact, he had graduated in 2003 from the University of Kansas in Lawrence where he obtained a degree in atmospheric science.

A meteorologist's principal tool for tracking a storm is Doppler radar, which shows the storm's progress in a color-coded matrix of pixels, green-blue to represent winds in the direction toward the radar, reddish-orange in the opposite direction. Meteorologists' data are supplemented by 3-D images generated by a software package called GR2 Analyst. On the night of May 4, these images showed cloud-like formations, which stalked the coordinate frame of Kiowa County like psychedelic renderings of Michelin Man. The time line for the storm's passage through south-central Kansas stretched from about 8 p.m. on May 4 into the early hours of May 5. The storm generated no fewer than nine tornadoes. The third was the Greensburg supercell. Three successive tornadoes, with EF3 status, later spread across parts of the county and adjoining regions to the northeast. The Greensburg tornado took shape around 9 p.m. southwest of the city, in Comanche County. It lasted for about one hour.

At 9:19 p.m., the Weather Bureau in Dodge City issued a tornado warning for Kiowa County and part of US Highway 54. The first of three telephone calls to Kiowa County had been made one minute before. The tornado was described as "extremely dangerous and life threatening." One minute later, Doppler radar showed satellite tornadoes were rotating around a major one—"multiple vortexes imbedded in the tornado" in the language of meteorology. The Office's GR2 Analyst software produced an image like a cliff to indicate the twister's strength and magnitude. By 9:28 p.m., the GR2 image had morphed into what looked like a giant fist with a finger pointing skyward. A second call was placed to Kiowa County. The Weather Service reported that the tornado warning would remain in effect until 10:00 p.m. for eastern Kiowa County. Highway US 183, which passes one mile to the west of Greensburg, was expected to be covered by debris. The tornado was approaching the highway about 7 miles south of Greensburg. Three minutes before, it had been reported 10 miles south of Greensburg and moving northeast at 25 mph. At 9:32 p.m., a giant vortex crossed US 183. The GR2 Analyst software showed what now looked like an alien spaceship hovering about parts of Kiowa County.

The third and final call from the Weather Service in Dodge City was made at 9:37 p.m. One minute before, Mike Umscheid had concluded, from viewing the latest Doppler radar image, that Greensburg itself lay directly in the path of the tornado. The GR2 Analyst now portrayed the tornado as a prehistoric creature, rising from the spaceship of the 9:32 p.m. graphic: a pterodactyl in the process of hatching. At 9:41 p.m., according to the *Signal*, "Umscheid pushed the button for a rarely used tornado emergency, used only when there is a certainty of a large, violent tornado about to strike a populated area."[30]

In Dodge City, it was 9:50 p.m. Umscheid watched as the tornado bore down on Greensburg. Four minutes were to elapse before the radar would show him what happened. "I was waiting to see what the next image would show," he said. "Will it die out before it hits town? Will it swing off to the east? Or will it continue to head north and go through town?"

****

The funnel of the twister was actually smaller when it tore through the Mennonite farms south of Greensburg, much smaller than when it hit the town itself. Still, destructive enough to wipe out houses, barns and workshops.

One couple to be hit hard by the tornado was the Goossens, Lloyd and Anita. A skilled carpenter, Lloyd Goossen has a reputation for the quality of his work. Although his wife, Anita, hails from Kansas, Lloyd emigrated from Manitoba, Canada, in 1977. The tornado demolished the Goossens' house and Lloyd's workshop where two clocks, one at the north

---

[30] Mark Anderson, "Meteorologist a local hero," *KCS*, May 7, 2008.

end of the building, the other at the south end, survived to record the exact times, seconds apart, of the tornado's sweep through the workshop. A level belonging to Goossen was hurled out of his workshop to a neighbor's field miles away. The markings on the level identified it as Goossen's. His neighbor returned it following the storm.

Lloyd Goossen's clocks

Lonnie and Terri McCollum had been following the progress of the storm on TV. The newscaster remarked that there was the possibility of bad weather, but the storm was tracking southeast of Greensburg. The worst that would come of it, or so the mayor thought, was that the small airport, east of town, would be wiped out. Still, he decided to open some of the windows in the house. When the electricity failed and the lights went out, the McCollums hunkered down in a corner of their basement, grabbing a mattress to cover themselves. At first it just sounded like a bad storm, but it got worse.

McCollum told his wife, "I think the house may be coming apart." When the storm subsided, McCollum groped through what remained of his cellar. With the aid of a flashlight, he made his way gingerly up a flight of debris-littered stairs. Once on the ground floor, he saw the night sky where his roof had been.

Alanna Goodman asked her father to drive her home when the rain began to fall. Before he had a chance to find his car keys, the tornado siren went off. Alanna reckoned that this occurred around 9:25 p.m. She grabbed her brother's cat, Spaz, and headed downstairs to the basement. Her own Siamese cat, Mocha, would soon perish back in Alanna's house. A foretaste of the tornado itself came about fifteen minutes later. Father and daughter first heard a banging sound, then a creaking noise, like the noise ships make when they are being pounded by waves. Glass began to shatter and furniture got shoved around. Dust and water poured down on them in the basement. That lasted about four minutes. Gary Goodman went out a side door to see how bad the damage was. Then the tornado spun around, sweeping over the town. This time it brought down the Goodmans' chimney. In trying to avoid being hit by his own chimney, Goodman twisted about and slashed his hand on some broken glass.

Looking for something warm to wear, Alanna went into her brother's bedroom. There she shoved her sandaled feet into Josh's prom shoes, grateful that her brother had such Brobdingnagian hooves. She also put on a pair of his camouflage pants and a baseball cap. Alanna and Gary climbed out of the basement, through a living room littered with broken glass. Exposed nails stuck out of the woodwork. The wind blew so hard that it tore Josh's cap off her head, whirling it into the night where it was lost forever.

There was no light, only lightning "like strobe lights." While she waited, her Dad checked on an elderly neighbor, Frances Nolan. Alanna felt helpless. At a loss to know what to do, she made her way over to a neighbor, Raymond Elliott, who lived just to the north of the Goodmans. Elliott was a construction worker who had spent the day painting in the town of Kinsley, 20 miles to the north of Greensburg. When he returned that night to the house he was renting, he was unaware that a tornado alert had been sounded.

Alanna needed help in getting onto Elliott's porch because the storm had taken away the steps leading up to it. In fact, Elliott's rental property had been moved about a foot off its foundation. Alanna encountered her parents' neighbor, still wearing his paint-splattered jeans and T-shirt. She asked him if he would help her. He noted that he had not met Alanna before. They shook hands and introduced themselves. It was all rather formal, given the circumstances.

From Frances Nolan's house, Gary Goodman yelled to Alanna, "Get a towel!" He needed to improvise a tourniquet to stanch bleeding in Frances' arm. Reluctant to make her way back to her parents' house, she turned to Elliott and begged him to get something—a towel, a T-shirt, *anything*—to stop the bleeding. At first, Elliott seemed not to comprehend what was being asked of him. Alanna repeated her request. Finally it registered on him what she wanted. After what seemed an eternity, he returned with some T-shirts which she asked him to take across the street to Frances Nolan's house. Minutes later, Gary Goodman backed his wife's Lincoln into the street. With Elliott's assistance, Goodman lay Frances on the back seat of the car. He drove off toward the hospital, but the engine died even before he could reach US 54. Alanna heard him cursing. She believed that half the town could hear him as well.

A day later, Josh Goodman, who was to graduate from Greensburg High School in a couple of weeks, rescued another family pet, a nanday conure parrot named after Bob Marley. In the confusion of the night of the tornado, the parrot had been forgotten. Josh found him the next day, sitting docilely and presumably in shock, beside the wreckage of his cage. Marley offered no resistance to being scooped up and taken away in Josh's pickup. A loquacious bird from a subspecies noted for its raucous squawk, Marley did not utter a word for the next two weeks.

The Goodmans' parrot, Marley

Charlie Jones had the reputation of being the best fisherman in Kiowa County. His wife, Pat, worked at the agricultural coop in Greensburg. Charlie's day job was in recycling. Over a period of five years, he claimed to have collected millions of tin cans in four counties of south-central Kansas, so many, in fact, that the company Charlie dealt with bought him a truck the color of Coors beer cans.

On the night of the tornado, when it became clear that they needed to find a storm shelter and fast, the Joneses and their 6' 3" son, Jesse, piled into the family's Ford pickup. About one block to the east of their destination, while driving down US 54, they saw the tornado wall coming right at them. Charlie had the truck in low gear, and when the tornado hit, he nailed the accelerator to the floor. Then he hit the brakes in a maneuver called "power-braking," to get as much traction on the road as he could. Just then, the Joneses' truck was rammed by a Ryder truck. The windows of the Ford imploded. Pat curlled up in a ball on the seat of the truck. "You could feel the rush of the wind," she said. She suffered head wounds and bled so profusely that her white blouse was stained crimson. "My brother in Minneapolis saw me on the Weather Channel that night," Pat said. "He commented on my red shirt."

Pat had been wedged between Charlie and her son. Jesse "had folded up like a giant taco," and that was fortunate, according to Pat, because otherwise he would probably have been killed, struck by a pole that came flying through the cab of the truck. For his part, Charlie had facial cuts and a broken finger. A stick of wood lodged in his throat, but Charlie managed to cough it out. With all the debris, mud, and dirt in the air, and with a ferocious wind to contend with, the only way Pat found that she could breathe was to wrap part of her coat over her mouth.

All three Joneses got out of the truck. The hail and the rain—"a cold, cold rain" in Pat's words—made it impossible to see one another as soon as they were separated by a few feet. Jesse yelled that he was going off on his own to see what help he could offer. The first thing he did was to pull the driver out of the Ryder truck. Jesse then disappeared into the night. Charlie screamed at Pat to follow him as he made his way up Bay Street, away from US 54, in the direction of a light he spied, a light that appeared to be coming from a long-deconsecrated church. There was no way Pat could do that. It was pitch black. She had taken her glasses off to wipe the blood from her face and could not see well. In her words:

> I felt the wind pick up real, real strong, like the tornado was coming back again the way some people said it did, and I said to myself, 'O my God, what am I going to do now? I don't know where Charlie is. I don't know where Jesse is.' The wind was getting stronger and stronger.

Pat held on to the railing of the truck bed. The wind was so fierce that she was lifted off her feet and suspended in mid-air, like a pennant, parallel to the railing. She clung to that railing for dear life. Then she was dumped unceremoniously into the bed of the truck. At that point she sustained a bizarre injury which she did not discover until three weeks later.

Charlie Jones eventually reached the former church. It was now serving as a house for a nurse practitioner, Jonell Sirois, and several kids. The roof of the building was starting to cave in, and Jones advised Sirois to get the hell out. At the time, he heard a voice telling him, "Everything is going to be all right. Just sit still."

Further on, he found another light, this time a flashlight. "Someone is giving me a tool," Charlie said to himself. He learned subsequently that the flashlight belonged to David Lyon, age 48, who lay dead a short distance away. No one realized that at the time because it was so dark. Charlie stayed about a block ahead of Pat, who by then was trailing him. She could see him by a rapid-fire series of lightning strokes. Finally she caught up to him. Charlie saw how she was bleeding. He wrapped her shoulders in a towel he found lying on the ground.

As Charlie Jones made his way east away from Bay Street, he heard someone "really cussing" about his pickup being totaled by the storm. Charlie walked over with the beam of his flashlight bobbing over the ground. There he found one of the more seriously injured victims, a woman named Colleen Panzer, who was in her late 70's. "I knew Colleen was hurt pretty bad because her right foot was on backwards," Charlie said. He asked Pat for the towel he had just given her. He then covered Colleen up with it. He told Pat that he would be right back. Then he made his way south to seek help and ran into firemen who were searching for people. They told Charlie, whom they addressed by name—"they knew me," Jones noticed calmly—that they couldn't go back with him because their orders were to do a house-to-house search. Charlie insisted. The elderly woman later died of her injuries. Charlie fretted about that. "What if I had hurried," he said. "But time sorta' stood still at that moment."

****

Megan Gardiner left the Lunch Box and sped home, arriving around 9:40 p.m. The speed limit was 30 mph, but she tore through town at 50-to-60 mph. "My heart was pumping like crazy," she said. "Luckily I didn't get stopped by a cop. I made it home with the tornado sirens still sounding."

Both Megan and her Dad Chris wrote about the night of May 4. An account of Megan's night first appeared in a collection of survivors' stories, *Greensburg–The Twisted Tales*[31]. The account given here expands on Megan's journal. Chris Gardiner's rendering of events is in the form of doggerel. At the beginning of his poem, he writes:

> We were worried about Megan getting home,
> Because we didn't want her to be alone.
>
> As she rushed through the door,
> She grabbed her lock box from the second floor.

---

[31] *Greensburg–The Twisted Tales*, compiled by Janice Haney, Mennonite Press, 2008.

That lock box, containing $1,000, was located in Megan's room. This was money that she had been saving for a trip to Italy which she and her brother, Matt, were to make that summer. When she got paid by the Lunch Box, Megan divided her wages into thirds, one third for daily expenses, one third for her account at Peoples Bank, and one third for her lock box. While Megan ran upstairs to retrieve the lock box, Chris welcomed into his basement a neighbor, Omero Carillo, his wife, Natalie, and their three children. When Megan got to the basement, she found the two girls, already in their pajamas, lying on the pillows she had made for TV viewing. The boy, Owen, in shorts and a T-shirt, was playing with the Gardiners' dogs, a boxer named Joe and a pug named Cheyenne. Megan engaged Owen in small talk. This was a ploy to keep him occupied. He told her that he "liked tornadoes but not tomatoes." In exploring the Gardiners' basement, he found a quarter which he stuck in his pocket.

A friend of the Carillos drove into their driveway, and Omero whistled him over. By now, quarter-sized hail was falling, and the wind had picked up. Everyone went down to the basement, where Megan watched the weather report on TV. The tornado was projected to hit Greensburg at 9:52 p.m. Omero and Chris Gardiner went out to collect hail samples. They were now the size of golf balls. Megan's cell phone read 9:49 p.m. With the sirens still wailing, she crouched by a couch, grabbing a pillow and a blanket to cover herself with. Then the power went out. All three of the Carillos' kids started to whine, and the Gardiners' boxer, Joe, began to shiver uncontrollably.

In Megan's words, "it was worse than diving deep under the water. . .probably one of the worst feelings I have ever felt. I thought that it was just me so I said, 'My ears are popping and hurt real bad.'"

Megan's cell phone read 9:50 p.m. She put her lock box by a doorway a couple of feet from the couch and stowed her purse underneath her. Then her world came apart. "It just got deathly silent. I mean this was freaky," Megan said. "I remembered that Kelsie Eck (a friend from Pratt who went through a tornado in 2002) said it got so still and quiet and I knew this was it!"

> Suddenly the windows blew in,
> We realized it was to begin.
>
> And as the house tore apart
> We had a sinking feeling in our heart.

Megan stuck her hand out to grab her lock box when the windows exploded. "They shattered into millions of pieces," she said. "I didn't see it but hearing it was enough."

By now, all three of the Gardiners were crouching in a row in front of the couch. The Carrillo family took shelter in a corner of the basement.

The next four minutes would likely remain etched in Megan's memory for the rest of her life. Her parents had long been aware of Megan's writing ability. In Junior High, she had written an imaginative description of food making its way through the digestive system. It was entitled "The Story of a Steak." Years later she had written "Toilet Bowl Story," a continuation of sorts of "The Story of a Steak." In her own words, this is what the little band in the basement of the Gardiners' house experienced.

> I heard the walls ripping off in pieces. Then something fell on my left shoulder and I covered my head with my hands. I decided to brace myself so I took my hands off my head and got down on all fours. The sound was like a jet going right over us, about to take off. While this was happening, my blouse started to fly up and I was thinking, 'O my gosh, I'm going to get glass up my back and I'm going to get cut.' But then something even heavier fell on my Mom. What happened was that an inside wall fell on her after coming over Dad, and she couldn't move. Dad said later that the wind tore loose the subflooring. Then it pulled boards through the anchor bolts on the stem walls.
>
> What seemed like a lifetime finally came to an end. I started to yell, 'Is everyone okay? Dad, are you okay?' He didn't answer so I kept yelling for him. Mom and I both yelled at him, and finally he answered that he was okay. Then he said, 'But, man, I've just been hit in the back by the dart board.' Mom was praying up a storm of her own. We stayed there for awhile because we weren't sure if it was over. Then I heard a man's voice say, 'Is everyone okay? I can get you out of here, but the stairs are blocked so we'll have to find a different way out.' I thought he was a first responder and I was thinking, now how long were we really down here if he is a first responder? But it turned out it was the Carrillos' friend. We all replied, 'Yeah, we're okay.' I managed to work my way out of the rubble.
>
> Dad had our boxer, Joe, on a leash pretty much the whole time. Mom and I were sheltering our pug Cheyenne. But when the windows exploded, Joe had jumped up and got away from Dad. I felt him brush against my leg, so I made a grab for him while I was under cover, but all I grabbed was debris. After the tornado passed and we got up, one of the first things I asked was, 'Where is Joe?' Cheyenne was fine, but we didn't know where Joe had gone. When I shone the flashlight around, I spotted Joe snuggled up on the couch where Dad had taken off

one of the cushions to cover himself up with. All you could see was his head because there were these boards on top of him.

I helped my Mom up, and when everyone made their way out of the rubble, I grabbed my purse and a flashlight. It was good that we had crouched down where we had because the couch was about the only thing in the basement not smothered by debris. The Carrillos had been lucky, too, because they had huddled in a corner: The angle in the walls, where Omero, Natalie, and their kids were crouching, had withstood the tornado. I was determined to find a way out. The whole basement was destroyed. I shone the light every which way to see how to get out. It took awhile, but I finally found a path. I crawled over plastic tubs filled with Christmas ornaments and blankets and duffel bags. We had called our basement our indoor garage. Finally I made it up by climbing on top of the porch roof which had come crashing down on the basement. Dad was the last one up. He used our basement shelves like a ladder. Just as he made it out, the shelves gave out.

When we crawled out of there,
A distinct smell was in the air.

As we looked around, no house could we see
But all the fallen trees and debris.

I looked around and pretty much the whole block was gone. Then Omero climbed up and handed me each of his kids, one at a time. Owen was the last kid, and he wouldn't let me go, so I had a kid hanging on to me while I tried to help Natalie and the other adults up.

The rain was turning to hail, the wind was coming up, and the lightning got more intense. 'Where do we go?' someone asked. We kept sweeping the piles of rubble with our meg flashlights. It seemed like we were trapped in our own yard beside the ruins of our garage. The debris was piled so high we couldn't find a way to get out. . . . Somehow we managed to get to the street. I saw that the north side of the high school was standing, but the south wing, directly opposite our house, had been levelled. We heard some voices down the street and shone our flashlights on people there. Brenda Hayes' house cattycorner to ours was still standing, or so we thought, and we started to go there for shelter, but as we got closer, we noticed that the roof was gone and we smelled gas. We yelled at the people inside—Brenda, her two sons, John and Cody, and Cody's fiancée, Maggie—to get out. They worked their way out a window, and someone asked again, 'Where do we go? We have to find shelter.' About then, a wind picked up big time. It blew me away, like when you walk a big dog and he takes off running and pulls you. The hail got so bad it was like being pelted by golf balls.

Chris Gardiner suggested that everyone head off to the Kiowa County Memorial Hospital, where he worked. That's what the Gardiners did, with Chris carrying Joe and Julie carrying Cheyenne. Brenda Hayes' group made its way north along Main Street.

****

Most of the residents in Greensburg who endured the tornado of May 4, 2007, were at home when it struck. But, like the Jones family, Jerry Diemart and his buddy, Andy Huckriede, were on the road when things started happening. They were luckier than the Joneses, however. They made it back to Andy's place where Huckriede made a beeline for his basement. Despite the juggernaut bearing down on him, Jerry decided to continue on to the J-hawk Best Western Motel, which his parents ran. They were away that Friday, attending the Sampler Festival in Garden City along with Erica Goodman and a lot of other Greensburg residents. Diemart did not realize until later that the bike he had just bought for his Mom had slipped off the bed of his father's truck. He slid into the motel parking lot, where his brother ran out to chivy him into the motel's storm shelter. This was the shelter that the Joneses had tried to reach but failed. The Diemart brothers got out the oil lamps stored on the ground floor. In the meantime, a police officer had managed to flag down as many as 100 people passing on the road. She shepherded them into the storm shelter. On their way there, the wind came up and pushed through the back door of the motel. Jerry helped the officer try to close it. Then the roof flew off. The ceiling of the first floor came crashing down, blocking the door to the basement and the storm shelter. Jerry and the officer took cover under a love seat in the lobby.

"It was like 'The Terminator,'" Jerry reminisced later. He heard the propane tank behind the motel blow up. It didn't take long to realize that Greensburg, which lost almost all of its trees' leaves, had become "the stump capital of Kansas." All the guests staying that night at the J-hawk Best Western made it safely to the storm shelter except for one woman using a walker, who occupied Room 125. She retreated to her bathroom. When the Diemart brothers emerged from their different places of refuge later that night, they found that every room in the motel had been destroyed with one exception: Room 125.

Remains of a Vintage Car

****

Like Lonnie McCollum, Representative McKinney thought that Greensburg would be spared, but when the sirens went off, his daughter, Lindy, insisted that they go down to the basement where they continued watching announcements on TV. McKinney tried phoning his parents on their nearby farm but to no avail. Then the Wichita TV channel they were watching solemnly announced that the approaching tornado "had the worst possible signature you can have on radar." At 9:40 p.m., the hail that Jerry Diemart had experienced in Haviland earlier that night started to pummel the town. The power failed but McKinney's cell phone was still working. He got a call from his neighbor, Chris Koss, who lived in a recently (and solidly) built brick house. Could his companion, Kelsey Schroth, and their baby son, Jayden, come over to the McKinney house? McKinney quickly agreed. "Tell them to hurry," he yelled and ran upstairs to the back door. There was no sign of either of them. Furniture was being whiplashed off the McKinney deck. McKinney retreated inside and began shining a flashlight to guide his neighbors to safety.

McKinney heard the glass blow out of the south side of his house and decided that the time had come to run back down the basement steps. He made it down four or five steps before the ceiling collapsed, hitting him on the head. Bleeding from a cut in his head, he dropped his flashlight but managed to get into the basement bathtub with his daughter, Lindy. Frantic over Kelsey Schroth's fate, McKinney was calmed by Lindy. She put her hand on his arm and started praying, first for their neighbors, then for themselves. The noise was deafening. They could hear bricks banging on the floor above. Things quieted down for a bit, then picked up as the tornado seemed to take a second pass over the town.

To get out of the basement, Dennis and Lindy had to clear debris from the basement stairs. They went outside to check the Koss-Schroth house. There was nothing left to it but the chimney beside a pile of rubble. "Anybody here?" McKinney cried out.

"Please get us out of here," Kelsey's voice came from a pile of rubble. "I have my baby with me."

McKinney yelled for help from two of his neighbors, High School Principal Randy Fulton and School Superintendent Darin Headrick, one of Lonnie McCollum's Rat Pack buddies. The three of them started digging with their bare hands through the rubble. Soon they uncovered a pair of legs amidst the debris. While Darin Headrick hauled away the dry wall that had collapsed, McKinney got the baby, Jayden, out. "His eyes were enormous," he said, "but there wasn't a scratch on him." Some more digging was required to free his mother who had been banged up. The family, McKinney learned, had taken shelter in a towel closet.[32] The tornado had lifted their car, which had been in the garage, to a point north of the house. An oily red fluid beside the car was spillage from its automatic transmission. McKinney's own pickup was carried half a mile away by the wind.

****

Again from Chris Gardiner's tornado poem:

> Lightning was flashing and rain turned to hail,
> The wind started blowing, then Meg stepped on a nail.

The Gardiners' group picked up stragglers as it made its way to the hospital, a little more than two blocks from their house. The wind was so strong that Megan had to squat down not to be blown away. Julie Gardiner was disoriented. Megan had to explain where they were. Only part of the hospital was still standing—the roof was off and part of the east wall was down—but Gardiner managed to unlock a door and get his group inside. The hospital dated back to the Cold War when bomb shelters were a common feature of public buildings. Once inside, Megan realized that she had been stepping on something. When she removed her shoe, she discovered that a shingle and a nail were imbedded in it. Chris and another nurse pulled these out. Neither had penetrated Megan's foot.

---

[32] Kelsey had been unable to get the door to the closet open. That was why she had not made it over to the McKinney house after Koss phoned.

> We arrived at the hospital cold and wet,
> And didn't know what was gonna happen yet.

Initially Chris Gardiner found only the four nurses on duty at the hospital with about fifteen patients who were badly stressed out. The basement was littered with debris, so the first thing Chris did was to enlist the able-bodied people to clear away what they could. The work proceeded by the jerkabout beam of flashlights. A woman arrived in the back of a pickup truck. All of the truck's windows and all its tires were blown out. The woman was suffering from internal bleeding. Gardiner strapped her onto a plywood board, held in place with someone's belt. A man with a broken femur was next to arrive. "I need your belts!" Gardiner bellowed, so that the man with the broken femur could be strapped to another sheet of plywood.

A friend of Megan's who worked at the hospital persuaded her, Julie Gardiner and the Carrillo family to go to the physical therapy room. Once there, the Carrillo children exchanged their rain-soaked clothes for hospital gowns. Megan got blankets to cover her shivering dogs, then returned to the hall where, by now, a lot more people had assembled. She asked people what they needed. They said, "Water." Megan's Dad advised her to give everyone only a shot glass of water. That way people wouldn't have to pee so much. In light of the lack of facilities, this turned out to be good advice.

When the Gardiners first arrived at the hospital, there was only one ambulance to be seen. Eventually, however, they began to stream in from all directions. Megan was surprised to see ambulances, not only from Pratt and Dodge City, but from the towns of Buffalo and Kinsley and from Comanche County.

The physical therapy room was filling up with children. Megan thought to herself—correctly as it turned out—that she would be babysitting all night. By the early hours of Saturday, she had convinced at least some of the kids that it was way past their bedtimes and that they should go to sleep.

Chris Gardiner carried down to the physical therapy room a man who was screaming with pain. Natalie Carrillo and Megan worried that the sight of the man's open leg wounds would upset the children, so they herded them to another part of the room that was curtained off.

Megan had to pee. Since the water and electricity were off, the toilets smelled like port-a-potties. She went twice that night at the rear of the hospital parking lot. When she got back to the physical therapy room the second time, she found that nurses were trying to get patients from the Behavioral Health Unit (BHU) to sleep on mattresses on the floor. The scene was something out of Dante's *Inferno*. In Megan's words:

The first guy had to poop. He was in a wheelchair, and the nurses couldn't get him into the bathroom. So they told him to hold it or go in his pants. He said 'no,' and began to argue with them. He ended up staying in his chair, and I don't know if he ever pooped or not. Then this crazy woman from the BHU with short, curly hair walked down the hall, cussing out the nurses and screaming, 'Let me go!' I didn't want her in with the kids. So I started to shut the door on her. A man grabbed the woman and told her that she should calm down. The crazy woman screamed back, 'Let go of me! You think you know how to handle a woman, but you're wrong.' As this was going on, I managed to slam the door shut.

They brought in some other patients to lie beside us, but they went right to sleep. So they didn't bother me. Most of the night I sat with my Mom, Natalie, the dogs and the kids, but I was still a little wound up and couldn't sleep even though I was tired. I was getting cold because I still had on my wet clothes. The nurses had offered me hospital scrubs, but I said no so that other people could have them. But when I ran my hands down my blouse, I felt bits of glass imbedded in it. I found a blanket and covered myself with that. Then I just leaned against the wall and let everything soak in.

While Megan was taking a breather in the physical therapy room, Chris had been busy. Early in the evening, a KDOT front-loader plowed through to the hospital. Chris asked the driver to clear a path to the highway. This made it easier for ambulances from neighboring towns to get to the hospital from US 54. Once at the highway, Chris ran into a paramedic who had a 4-wheel drive car. Chris persuaded him to drive to Dillons Supermarket which had been set up as a triage center. There Chris saw a doctor he recognized, the only person apparently with a working cell phone in Greensburg. He used it to call his son, Matt, in Wichita.

Natalie Carrillo's sister from Dodge City came to take the Carrillos home with her. Before leaving, Owen produced the quarter he had found in the Gardiners' basement. "Look," he said, "I still have it."

> Working thru the night, there seemed to be no end,
> It was a good sight to see our camping friends.

Megan lost track of time and was talking to Julie when "all of a sudden all of our camping buddies walked in." Her journal entry reads as follows:

I think my jaw actually dropped when I saw them. Terry Reno, Doug Freund, and Mark Richard. Then my Mom turned to see what I was looking at. She gets up and starts to cry while she was hugging them all. She asked if they were camping and they said, 'Yes, and we tried to call you guys and couldn't get through. We couldn't wait any longer so we said 'Screw it!', hopped in the truck and headed here. With all the debris on the roads, we had to go cross-country. It took 90 minutes to get to Greensburg from the lake.'

Not very long after that,
In walk Don, Marge and Matt.

With all the hugs and tears,
It relieved their worst fears.

Julie, Megan and the dogs went to the lake,
While a difference the rest of us tried to make.

The people mentioned in the first line of this part of Chris' poem were Megan's Aunt Marge, her Uncle Don, and her brother, Matt, who came all the way from Wichita when he heard from Chris of the destruction at Greensburg. As Chris had requested, the Gardiners' camping buddies drove Julie, Megan, and the dogs to the family's camper outside Coldwater, in Comanche County. Before leaving town, Megan asked her friends to drive her past her house so that she could retrieve her lock box. They agreed, but she did not find it.

En route to Coldwater, Megan saw dead cows beside the road. She and Julie and their friends passed a gas plant that smelled of gas, so Megan called 911 to alert the authorities to a gas leak. At the camp site, Megan and Julie showered. Megan discovered that she had debris in her hair.

When I took out my messy bun, all sorts of crap fell out: Bits of shingles and insulation and glass. I had a cut on my head that nearly killed me when soap or shampoo got in it.

Try as she would, she couldn't sleep. Chris hitched a ride down to Coldwater with his son, Matt. In his own words:

Chris got in the shower and scrubbed his ass,
Where he found a whole bunch of glass.

At the time of the tornado, he had been squatting down with his head buried in a pillow. Shards of glass from shattered windows had imbedded themselves in his butt.

****

When the coast was clear, Lonnie McCollum made his way down what had been the Main Street of Greensburg. The Baptist Church was gone, and firemen who came up to him said "the whole town looks like this." Brick walls had collapsed on vehicles at the intersection of Main Street and US 54 (Kansas Avenue). His wife, Terri, had hurt her knee getting out of their basement. They were separated until 4 a.m. Saturday, when McCollum found her, together with other townsfolk who had been injured, at the Kansas Department of Transportation (KDOT) building. People were wandering about like lost souls, without shoes or glasses, and McCollum was speechless with awe: Think, he thought to himself, Greensburg, after the tornado, looks like Hiroshima, after the bomb. It was determined that the Greensburg tornado had a diameter of 1.7 miles.

At that moment, in the wee hours of Saturday, May 5, 2007, when Lonnie McCollum, the environmentalist mayor—his wife injured, his house destroyed, his collection of Armani ceramics mostly in smithereens, his idyllic, Norman-Rockwell town of Greensburg in ruins—it was at that moment McCollum resolved to rebuild the town green.

VOLUNTEERS

"I do not know where Kansas is, for I have never heard that country mentioned before. But tell me, is it a civilized country?"

"Oh, yes," replied Dorothy.

—Wonderful Wizard of Oz

Luke McFadden suffers from spina bifida, a congenital defect in which the spinal column is imperfectly closed. The way Luke walks—striding with one leg, then dragging the other behind—gives away his disability. Disability be damned. Luke is one gutsy guy. The man who was to succeed Lonnie McCollum as mayor, John Janssen, related an anecdote to show this: Luke had been helping Janssen with his cattle. His horse stepped on a hot wire and streaked off, throwing Luke to the ground. After he picked himself up, Luke climbed back in the saddle and continued herding Janssen's cattle. Yup. Luke is one gutsy guy.

Like his younger brother, C.J., Luke works as a volunteer fireman for the nearby town of Mullenville. He has done so since 2004. C.J. has logged three years as a fireman. Luke and five of his mates came by the Mullenville fire station, one year after the tornado. Two of them were related: Cody Sherer is the son of Rick Sherer. Another of the fire fighters there was Danny Toner, a big man who looked as if he could cope with disasters. It was Danny who had organized this session at the fire station.

At 26, Luke is among the younger fire fighters in Mullenville—he is, in fact, the third youngest of the ten volunteer firemen in the town. He is treated, on occasion, as the group's mascot, but on the night of May 4, 2007, Luke played as heroic a role as his fellow fire fighters who came to Greensburg on a search-and-save mission. The Mullenville fire fighters were among the group of first responders who, as Lonnie McCollum said, "saved the day for Greensburg."

All the men carry pagers so they know when to report for duty. On the night of the tornado, they had gathered first at 6:30 p.m. Then once again at 9 p.m. They left for Greensburg half an hour later, driving down in two fire engines and a couple of private cars. They parked in the passing lane of US 54, just to the west of the town. The time was around 10 p.m. By then, the traffic was backed up on the highway for miles. Around 10:15 p.m., Cody Sherer got a call from the county fire chief. The fire fighters were given permission to enter the town. They were equipped with crow bars, one chain saw—which they never used—and a paltry number of flashlights. Cody and one of his comrades proceeded three blocks into town. What they found there dumbfounded them.

Houses were gone. Telephone poles lay across the road. At the John Deere dealership, overturned combines were little more than stacks of twisted steel. Flashes of lightning revealed the devastation all around them. C.J. McFadden noticed that the Lunch Box, where Megan Gardiner had washed dishes earlier that night, had been lifted off its concrete pad. The only reference points were Dillons Supermarket on US 54 and the shell of the Greensburg High School further south on Main Street. After the Mullenville High School had been closed, Luke, C.J., and Cody Sherer had all gone to Greensburg High. As the fire fighters made their way along US 54, they encountered a string of bizarre scenes. Rick Sherer remembers seeing a horse, stranded and motionless in the middle of the street, apparently in shock.

The Mullenville fire fighters had had little training in handling a catastrophe on this scale, although all of them had been enrolled in a first responders course. This had started the previous January. Ironically, it was to have ended in mid-May. The Greensburg tornado was like the lab part of the course. Most of the men felt that they should all have been given passing marks on the strength of what they did on May 4 and the days that followed.

The firemen split up into teams of two and began a methodical search of the town, street by street. Charlie Jones had cajoled one of these teams into helping rescue Colleen Panzer. The fire fighters' only light, apart from their flashlights, was generated by a portable generator. Names of survivors in good physical condition were written down and sent back to the triage center at Dillons.

Greensburg was a mess. One article to appear in the *New York Times* (*NYT*) on Sunday, May 6, 2007, captured the desolation of the town:

> For perhaps 80 square blocks, old brick buildings and new frame ones were flattened into twisted piles of masonry and timbers. Attics tipped to the street, upstairs bathrooms were open to

the sky, roofs were flush with the ground, and cars perched precariously on heaps of debris. . . .

The city administrator, Steve Hewitt, rode out the twister in his basement with his small son. After the tornado, Mr. Hewitt said he "went upstairs" only to find "there was no upstairs."[33]

A few days later, a valedictory to familiar landmarks on Main Street appeared in the same paper:

> Hunter Drugs is gone, with its ruffled blue-and-white fringe of an awning, its walnut and marble booths, its red leatherette and chrome stools perched before the old-fashioned soda fountain. So is the Twilight Theatre, with its pressed-tin ceilings that date from the time it showed silent pictures. [34]

The Mullenville fire fighters returned to their homes around 6 a.m., Saturday, May 5. After a shower, a change of clothes, and breakfast, they headed back to Greensburg. Four of the men—Cody, Danny, Luke, and C.J.—stayed in Greensburg for the next nine days, 24 hours a day. At first, they were lodged east of the town. Then a temporary firehouse, where they slept on cots, was set up at Apache Services, a truck repair and maintenance facility north of the golf course. They ate at the Salvation Army tent. Cody and Luke remembered the day they were served a meal of pizza, a welcome change to the routine of those nine days.

In those early days, things were confused. There was a lack of coordination among the various organizations on the ground. Luke became adept at bypassing no-go zones set up by the National Guard. One of the saddest memories the men had of their early hours in Greensburg was to hear that someone at the Command Center—it was not clear who—had called the Mass Casualty Recovery Unity in Fort Hays to say that "we'll need 500 body bags" for the dead in Greensburg. That turned out to be a wild overestimate. There were only 11 fatalities in Kiowa County as a result of the tornado. Still, those fatalities included people the men from Mullenville knew personally. One such victim was Claude Hopkins, a well-known resident of Greensburg who had ridden as a Patriot Guard for the American Legion. His body was found one block north of his house.

The tornado had come from Comanche County to the south. A spotter from that county, Greg Ellis, started to follow the storm on Friday at about 6:30 p.m. Ellis was the fire chief of the town of Coldwater where the

---

[33] Adam Nossiter, "An Empty Place Where a Kansas Town Once Stood," *NYT*, May 7, 2007.
[34] Pam Belluck, "Welcome to Start From Scratch, U.S.A.," *NYT*, May 13, 2007.

Gardiners had sought shelter on May 5. A heavy-set man with a greying goatee, the fire chief had held his job in Coldwater for the previous 11 years. His second cousin was Representative Dennis McKinney. On the night of the tornado, Greg Ellis had come within half a mile of the twister. At that time, the storm had rotational velocities of 80-to-90 mph. Even with these relatively modest speeds, it had been impossible to open the door of his vehicle.

Just inside the Kiowa County line, a cottonwood tree came crashing down in front of him. Ellis backtracked and started checking farm houses on an alternative route to Greensburg. Halfway there, he learned that the tornado had wiped out the town. Ellis paged for more units to go north. Eventually he was able to get to the town, where a bulldozer made a path for his fire truck and for the ambulance that accompanied it. One snapshot of the devastation to linger in Ellis' mind was that of a two-story house squatting in the middle of a street. Ellis recalls in particular that the state troopers on-site were failing miserably to direct traffic on US 54.

"The Weather Service needs eyes to verify what they're seeing on their radar," said Ellis. Spotters report on the size of the hail they encounter, the straight line speed and direction of the tornado and the wall cloud which typically comes behind the rain to the southwest of the tornado. Hail can achieve softball size. Even bigger. Ellis said that it was possible to cut a large hailstone in half and expose its rings, a bit like the rings of a tree trunk. They tell you how many times the hailstone has been wafted up and down during a storm before it finally falls to earth.

A local employee of the Kansas Department of Emergency Management (KDEM) met with the fire chief of Greensburg immediately following the tornado, in front of what was left of Dillons Supermarket. The two men asked the emergency manager of Pratt County, who was sitting in his own car just outside Greensburg, to oversee emergency management while they went to the KDOT building to set up an Emergency Operations Center. Some KDEM official—it was unclear who—called headquarters in Topeka and asked that the request for an emergency declaration be forwarded to the office of Governor Kathleen Sebelius. The governor forwarded the request to federal officials in Washington. By Sunday, May 6, 2007, President Bush had issued a disaster declaration for the Kansas counties affected by the tornado. This permitted the Federal Emergency Management Agency (FEMA) to come in.

****

On May 5, 2007, the Gardiners ate breakfast at a friend's camper. There they watched TV and learned that more severe storms were predicted for their part of Kansas. The Gardiner clan decided to drive to Pratt where they could stay with friends. But the National Guard would not let them pass through Greensburg. "Sorry, sir," Chris was told (erroneously) by a Guardsman that "martial law is in effect, and no one can enter."

The Gardiners took country roads to get to their friends in Pratt. Once there, they watched TV news. Said Megan:

> Of course, they wouldn't let Greensburg residents enter [the town] but they did let every news channel in the country in to take pictures. To me, that doesn't make any sense at all.

Lonnie McCollum himself lived for a couple of days in a pickup truck and did without sleep. He and City Manager Steve Hewitt split up their responsibilities: Hewitt took care of business while McCollum handled people's problems.

Pat and Charlie Jones were taken to Dodge City where Charlie spent three quarters of an hour in the emergency room of the local hospital. He needed stitches for cuts caused by the implosion of his windshield. He marvelled over the fact that he wasn't dead. Drenched through and through, Charlie caught cold from the AC, first in the vehicles which took him to Dodge, then from the AC in the hospital itself. "I damn near got pneumonia that night," he said. The family sought refuge at Charlie's brother's place in Dodge. Then they came back to Greensburg to retrieve Charlie's guns and his fishing rods. The Joneses' house was gone, and all their vehicles were demolished. It wasn't until three weeks later that Pat realized that she was suffering from an infection in her left leg. A surgeon in Dodge City cut into her while she looked on. "Charlie couldn't believe that I'd watch as he did that," Pat giggled. The surgeon dug out—of all things—a kernel of corn. "Once he got it out, the leg healed a lot faster," Pat said. She kept the kernel of corn as a souvenir of her tornado experience.

Jerry Diemart had spent the first six hours after the tornado on a search and rescue mission of his own. One of the people to benefit from his Good Samaritanism was Frances Nolan, the 96-year-old woman who had lacerated her arm. The tourniquet which Gary Goodman had applied earlier that night was still in place. But Frances continued to bleed profusely all over the back seat of Erica Goodman's Lincoln. Jerry's nephew, Corey, offered her his shoes, but she was too weak to walk. So Jerry "potato-sacked" her and got her to another neighbor's car. This got them to the first obstruction in the road at which point Jerry transferred her to a pickup that transported her to US 54. First responders from nearby towns took her from there.

Also on their list was a relative—Aunty Evon—who lived on Bay St. Jerry and Corey ran to her house to find it pancaked. (She was safe in her basement.) Cars were wrapped around telephone poles like fat slabs of licorice. There was a humongous puddle, two feet deep, created when the town water tower, adjacent to the Big Well, collapsed. One of several damaged vehicles was a truck that belonged to Jerry's Dad. The windows were blown out and one of the tires was flat, but Jerry got it started. At daybreak, the National Guard warned him that he was either leaving town or going to the pokey. But Jerry is nothing if not pig-headed. He stayed in the vicinity of Greensburg, successfully evading the authorities.

A day after the tornado, Jerry Diemart and his nephew were barrelling down a dirt road called the Cannonball. It runs parallel to and one mile north of US 54. Corey was feeling cocky. He said that he was scared of "no ole tornado." Talk of the Devil and he'll appear: As if on cue, a mini-tornado whirled out of nowhere and bore down on the car. Corey was working up a cold sweat. He decided to outrun the twister. Even when they changed course, abandoning the Cannonball to yaw down the extension of Main Street which led to Jerry's parents' farm, the funnel cloud kept sucking at the rear bumper of their Chevy. Corey plowed through the tornado-induced lakes covering the roadway. The last of these killed his transmission. Still that ole tornado came after them, grinning from ear to windy ear. What was a Kansas boy to do? The one gear still working was reverse, so somehow Corey spun the car around, revved up the engine and screeched off in reverse. He and his uncle stared through the windscreen into the funnel cloud hot on their heels. Visually it was the stuff of a Keystone Kops comedy. Corey Diemart made no more smart-ass remarks about tornadoes that day.

Mother Nature may have her own way of playing practical jokes on human beings, and humans have theirs: One prankster managed to slip a pair of red slippers—the perpetrator had seen the movie of *The Wizard of Oz*—under a corner of Jerry Diemart's house which had slipped off its foundation. Incredibly Andy Huckriede found the bike which Jerry had bought for his Mom and managed to salvage it.

Most aspects of the tornado strike were not so funny. It cost eleven people their lives. Harold Schmidt lost his house and his life when the twister picked up his car and dropped it on him, crushing him to death in the basement of his own house. A passerby from Albuquerque, New Mexico, was sucked through the window of his camper. This was the only non-resident fatality to occur in Greensburg. The man's body was found in Kiowa County State Fishing Lake, just to the northwest of town.

At the John Deere dealership, one of the firm's combines was carried 25 to 30 yards from the agency lot to a holding pond on company property. Some combine platforms, weighing as much as 6000 pounds, were carried about one quarter to a half mile north to the Kiowa County State Lake where the body of the Albuquerque man had been retrieved. "The damage done by the tornado to our dealership exceeded $18 million," Mike Estes, part owner of the dealership, wrote in an e-mail. "This was the single largest claim in our insurance carrier's history."

The Gardiners had been prevented from entering Greensburg on Saturday, May 5, but the next day Chris was requested to return to the hospital to help remove materials before they were water-damaged. Once again, from Megan's journal:

> When we pulled into town, it started to rain hard. I was worried that another tornado might come. I just couldn't believe the destruction that we saw. It's nothing like the pictures and TV. In real life, it's one hundred times worse. We made it to the hospital and it was still raining, even harder. We went downstairs and cleaned off the shelves. . . . Dad got a call from his sister, who said that there was a tornado heading straight for Greensburg again. I was like, 'Great, just what we need.' We were in the basement and safe, and there were a lot of mattresses people had slept on the night before. I was going to pull them over me if anything hit. But nothing ever came, just a little hail and lots of rain.

There was one sliver of consolation: On the Monday when the Gardiners returned to Greensburg, Megan found her lock box. She and Matt went off to Europe in the summer of 2007, spending ten days in Greece and Italy.

****

The most extraordinary group of volunteers to go to Greensburg involved a kernel of young political activists. Known as Kansas Mutual Aid (KMA), this Lawrence-based anarchist collective maintains community gardens in Lawrence. It participates in the occasional anti-war demonstration. This has brought them to the attention of the local police. The KMA made a series of trips to Greensburg, starting on May 12, 2007, the Saturday after the tornado blew the town apart. These continued every Saturday for the rest of the month and into June.

On the first of these trips, four KMA members drove down to Greensburg in a car that ran on vegetable oil. Two of them, Dave Strano and Joe Carr, wrote about their experiences that day. Strano also wrote about later trips. So did two other friends, Amber Fraley, who did not go to Greensburg herself, and Jordan Ferrand-Sapsis, who did.

Dave Strano is an army brat. He was born in Germany where his father was in the US Army. In the spring of 2007, Strano was working in Lawrence as a school bus driver. He, Joe Carr, and two of their friends went to Greensburg where AmeriCorps had set up a base of operations to coordinate volunteer efforts.[35]

Dave Strano

On May 12, 2007, the anarchists reported to AmeriCorps' red-and-white tent There they strapped on the red wristbands that identified them as aid workers. Besides their intent to provide assistance with recovery, the KMA had come down to assess the situation in Greensburg. Strano described his group as a "fact-finding delegation." One of their interests was the condition of prisoners in the county jail, both during and after the storm. The issue of prisoner treatment turned out, not surprisingly, to be a non-starter since there was usually no more than one person in the Greensburg jail at any one time. The nighttime jailer was a single mother.

The KMA members spent several hours hauling damaged items from the basement of one family's house. The owners, according to Joe Carr, were among the fortunate since they were fully insured and qualified for an SBA loan. Their basement, recently remodelled for $20,000, had been flooded after the tornado. Still, it had provided shelter not only for the owners but for four of their neighbors. The KMA's efforts were appreciated. "Next time, bring fifty more with you," said one of the Greensburgers.[36]

---

[35] This federal agency stayed in Greensburg from May 5 to September 15, 2007. Over the summer months of 2007, it oversaw the work of more than seven thousand unaffiliated volunteers, including the KMA cadres.

[36] Material about the KMA comes from several sources. Dave Strano and Joe Carr gave interviews. Both men did some writing of their own about their experiences in Greensburg. Strano's article, from which this quote comes, "Somewhere over the

Carr observed that the KMA car was shadowed by an Overland Park police vehicle for more than one hour. (Overland Park is a suburb of Kansas City.) The police even followed Carr and his companions as they walked down the debris-littered streets of Greensburg. The anarchists were particularly sensitive to being tailed because they are wary of "the man," given their history of run-ins with authority. Anarchists like Joe Carr were quick to link what they saw in Greensburg with developments elsewhere, even to conflicts in the Near East. Carr had spent time working there as a volunteer. Whippet-thin and pony-tailed, he radiates a kind of high-strung nervous energy. Carr had had more adventures in the previous five years (he was then 27) than most men pack into a lifetime.

Joe Carr

The young firebrand first learned about anarchism in college. A native of Kansas City, Missouri, he attended Evergreen State College in Washington State where he was a classmate and friend of Rachel Corrie.[37] In 2003, both of them made their way independently to the Gaza Strip, then occupied by Israel. They worked there under the auspices of the International Solidarity Movement (ISM), a Palestinian-led anarchist group. The ISM's stated mission was to repair the region's crumbling infrastructure and to prevent the destruction of houses and olive groves by the Israeli authorities.

Rainbow: KMA Report from Greensburg, Kansas," can be found at *Infoshop News*, May 14, 2007; Joe Carr's "Disasters of our Age: From Fallujah to Greensburg," at *www.lawrencesolidarity.org*, May 12, 2007.

[37] The September-October issue of *Sierra* listed Evergreen as Number 3 among its 100 greenest schools in the country.

One measure the group adopted was to stay in houses slated for demolition. In one well-publicized confrontation, Corrie was crushed to death by an Israeli bulldozer as she stood in front of one such house. This occurred two months after Carr had arrived. Another of his friends, a Brit named Tom Hurndall, was shot by an Israeli soldier. Hurndall succumbed to his injuries after lingering in a coma for eight months. Traumatized by these deaths, Carr returned to the US where he underwent therapy.

Carr went back to the Middle East, this time to Iraq, as a member of the Christian Peacemaker Teams (CPT). He stayed during May 2005, mostly in Baghdad although he spent a few days in Kabala and visited the battle-shredded city of Fallujah. Carr eventually produced a CD, *Resistance to Empire*, of hip-hop music inspired by his experiences in the Middle East. The second song on that CD is entitled *Fallujah*, and it contains the lyrics:

> Fallujah, my eyes cried to drive through ya'
> 'cause you were hammered, clawed, and sawed as they screwed ya'.
>
> And they shot you full of holes like Swiss cheese or Gouda,
> But you insisted on resisting, and I raise my fist to ya'.

On his only trip to Greensburg, May 12, 2007, Carr was struck by how few residents of the town were actually there. He also noted how the National Guard had systematically searched houses to determine whether they could be rehabilitated, a process during which the Guard was reputed—incorrectly as it turned out—to have seized residents' weapons. Joe said that a kind of code had been spray-painted on some houses to indicate what the Guard had confiscated there. "SO," for example, stood for "sawed-off shotgun."[38]

The confiscation of guns was an issue which forged a bond between the KMA and Greensburg residents. This might appear paradoxical, given the political views of Greesnburgers. For the KMA, however, possession of firearms goes to the heart of state repression. The residents of Greensburg would be unlikely to use anarchist language to express themselves, but a belief in the right of private citizens to own firearms dovetailed seamlessly with the anarchists' philosophy of local control. Joe Carr, interestingly, felt that people in Greensburg were more in tune with anarchist ideals than were the citizens of cosmopolitan, university-centered Lawrence.

In an article she wrote for a weekly Lawrence paper, Amber Fraley cited the case of Larry Brennan, who said that he could understand why

---

[38]Major General Tod Bunting of the National Guard denies that the Guard seized any weapons other than those lying in the road "where children might pick them up." He points out that it would not be possible to search nonexistent houses for weapons.

some people opposed the private ownership of guns and how, in a disaster situation like Greensburg, unsecured guns should be in the possession of the appropriate authorities.[39] But his guns, Brennan contends, were locked up. When he was permitted to return to his home, he found that his gun case had been broken into and his firearms taken away. "They don't have the right to break into the remains of your house and take your guns," Fraley quoted Brennan as saying. "They went house-to-house, room-to-room, closet-to-closet, even in structures that were still standing, for three days before they let anybody back into town."

Along with other Greensburg residents, Brennan called the National Rifle Association (NRA). "They were here the next day from Washington, D.C.," he said. And they brought a film crew with them.

The NRA wired Brennan with a mic and shadowed him as he and other townies sought to have their firearms returned. After three days, they were permitted to enter a 40-foot trailer which was piled to the roof with unlabelled weapons.

> I asked, 'Which ones are mine?' 'We don't know,' they told me. You sifted through rows and rows of terribly rusty, mostly unrecognizable guns. I said, 'I could take this $1200 Browning over my $800 Remington?' And they just shrugged their shoulders and said, 'Yeah, you could.'

In a press release posted on their web site, the NRA stated:

> Last week, NRA reported that we were investigating allegations of gun confiscation in the aftermath of the tornado that ravaged Greensburg, Kansas. . . .After investigating these complaints, there was no evidence of any illegal gun confiscations or seizures.

Aaron Einsel is slight of build, a man with a moustache that conceals a scar incurred years before. When interviewed, he wore sun glasses and a baseball cap with "First National Bank of Pratt" scrolling above the visor. Einsel had sold guns out of his house, a white farmhouse with neat blue trim, north of Greensburg until "Paul Morrison [then Attorney General of Kansas] shut me down."

Einsel, understandably, was vehement in his opposition to the authorities' confiscation of firearms, which he looked upon as a violation of his and his neighbors' constitutional rights. "Can you quote the Second Amendment to the Constitution?" he asked. "Just 21 words." The confiscation

---

[39] Amber Fraley, "Guns and Anarchy in Greensburg," *The Lawrencian*, June 1, 2007.

of residents' firearms had prompted the introduction in the Kansas Legislature of a law to protect gun owners' rights. This was House Bill 2811. It prohibits state officials from seizing, temporarily or permanently, legally owned firearms or requiring registration of any firearm "for which registration is not required by state law." Einsel had driven up to Topeka to testify on behalf of the bill.[40] "Kansas is one of the most self-sustaining states in the union," he said. He went on to cite an anecdote which everyone in Greensburg seemed to know a version of:

> Ask Jeremy Butler to tell you the story of his mother's telling law enforcement officials to take a hike. No one was going to take her shotgun away. Twice they came to her door, and she refused. The third time they came back with someone from the KBI (Kansas Bureau of Investigation). Still she stood her ground, and she weighs 98 pounds in wet clothes.

****

The modest stick-frame house in which Shirley and Rex Butler live sustained some damage during the tornado, but unlike structures around them, it was left standing. Both Butlers come from small towns, and they are self-reliant people. "Teach your kids how to grow things," Rex advised during a walk around his organic vegetable garden behind the house. He earned his living by holding down a part-time job at the post office and working in neighbors' gardens.

The day after the tornado, when law enforcement officials began to comb the streets of Greensburg, the Butlers offered them the use of their generator to run their radios. Their offer was declined. An offer to help with the search and rescue mission was similarly rebuffed on the grounds that the work was too stressful for civilians—even civilians like Rex who had done a two-year stint in the Marine Corps.

The Butlers had their own version of the story about their confrontation with the police. They said that on the Sunday following the tornado, while they were having breakfast on their front porch, the police came by and ordered them to leave town. They were told—incorrectly—that Greensburg was under martial law. The Butlers were determined not to leave their house and said so. The cops went away but came back to reissue an order to vacate. "They had an intimidating stance," Shirley Butler said, "and they were shouting." The way she retold the incident, the cops kept their hands close to their revolvers, like cowboys in an old-fashioned Western

---

[40]HB 2811 was subsumed in another piece of legislation which was passed and codified in the statutory code of the State of Kansas as KSA 48-959. It became law on July 1, 2008.

shoot-out. Rex Butler advised the police that if they tried to go through their front door, "things are going to get noisy."

The police went away, and the Butlers decided to maintain a low profile not to attract undue attention. They had a 10,000-watt generator which cost around $800 per month to operate. So they had power—as long as friends kept them supplied with gasoline—and they could eat off what they had on hand, or could get, again from friends, or harvest from their garden.

Years after the event, Lonnie McCollum talked about police behavior in Greensburg during the first few days after the tornado. In his opinion, local police units leaned over backwards to be helpful. But others from further afield "used Gestapo tactics." This revolted the ex-state cop who blamed much of the problem on the Wichita Police Department. He mentioned one incident in which an elderly couple, driving into town, was stopped by police, the lights on their patrol car flashing. Two patrolmen got out of their car and approached their quarry—and here McCollum mimicked them the way Shirley Butler described her own brush with out-of-town cops—like gunslingers in a John Wayne movie.

One of Shirley Butler's shotguns

Shirley's two sawed-off shotguns were lying in the living room. She propped one of them on the front porch so that it could be photographed. Although she had never threatened the cops with it, her son Jeremy's mother-in-law gave the Butlers a mock traffic plaque, which read "Caution Crazy Woman with Shotgun." Shirley had the sign taped on her front window. Rex had bought the shotguns from Aaron Einsel who acquired them from the Greensburg police in exchange for some M-16s. Rex said that people kept firearms as heirlooms. The region around Greensburg had some of the best deer and quail hunting in the state, if not the nation.

****

Dave Strano returned to Greensburg on May 19, 2007, with four other anarchists. One of them was Jordan Ferrand-Sapsis. They were stopped at a checkpoint outside town where they were given a pass to display on the windshield of their car. Their goal was to contact Christian Disaster Relief, a group with a reputation for effectiveness and savvy, to see if they could use land adjacent to the Mennonite Church to set up a sort of clearing house for material and volunteers coming down on subsequent trips. The church itself had been levelled by the tornado. It had been replaced by a trailer at the time of the anarchists' visit. According to Strano, the anarchists were warmly greeted by the Mennonites, but they were told that there was no room available for them to use.

Continuing down the road, the five young people encountered a man who seemed interested in letting them use his property, which was close to Dennis McKinney's house. While they were talking to him, a police car pulled up. A sheriff from Dickinson County told them that they would need FEMA approval to set up a base of operations. The Lawrence anarchists returned to the center of Greensburg where two of them went to talk to FEMA officials. The remaining three, including Dave Strano, went to grab a sandwich and a glass of water at the Salvation Army tent.

After a brief meeting at the Kiowa County Emergency Response Command Post, the two young anarchists returned to find Strano and their other two comrades in a tense stand-off with a cop from the Olathe Police Department. (Olathe is a suburb of Kansas City.) The policeman advised the anarchists to "keep your hands out of your pockets where I can see them," according to Ferrand-Sapsis. The group of five was moved away from the mass of volunteers to a side street because, as the Olathe cop said, he didn't want a scene.

More police showed up, as well as a female officer of the KBI and representatives of Homeland Security. Strano was separated a few feet from his comrades and told to leave and not return. If they returned, they would be arrested. To use military jargon, 'securing' the anarchists would be an unacceptable drain on police resources.

Ferrand-Sapsis was not surprised that the anarchists had been recognized by the police. Many of the group had been arrested in the course of their anti-war and anti-recruitment protests in Lawrence. The police accompanied the anarchists back to their vegetable-grease-burning car, which was escorted out of town by several police cars, their lights flashing.

The mercurial Joe Carr gets the last word here. Once again, from his hip-hop song, *Fallujah*:

> Occupation forces roam through the streets in tight groups
> Shia from the south, contractors and US troops
>
> Mob cowboys, thugs rarin', blarin' their guns
> Boys in hummers and pickups just having too much fun
>
> You drive too close and get blown away
> Better pray and hope they're not having a bad day.

****

In September 2007, during the high school homecoming parade, Chris Gardiner—who was president of the school booster club—shared a float with the meteorologist, Mike Umscheid. Umscheid was Grandmaster of the Parade. During the course of his visit, Umscheid learned of Megan's journal. He asked to see it and was so bowled over by its power and accuracy that he borrowed a copy. He subsequently incorporated parts of Megan's account of the tornado into a technical presentation which he made, first at a meteorology conference and later in Greensburg, on the anniversary of the tornado. He cited Megan's journal, as well, in a paper, entitled:

The Greensburg, Kansas Tornadic Storm: A Storm of Extremes

That by now well-cited journal concluded as follows:

> People say, 'Oh, I'm so sorry. I know what you are going through.' Actually they don't. You have no idea until it happens. We were really sore for at least one week after it happened because of everything that fell on top of us. That's the kind of stuff people don't understand. If you don't think anything is exciting in your town, and you live in Tornado Alley, then believe me, that could change in about four minutes. That's about how long the tornado lasted. I never want to experience that again. One is enough, one too many for me. It turns your whole life upside down. Now I get very paranoid when a storm comes. Even if it isn't bad, my memory flashes back to that night. May 4 changed the lives of 1500 people when the EF-5 tornado came through. Only 1500 people can say they were in an EF-5 tornado. I was one of them.

KATRINA'S GHOST

The Witch was too much afraid of the dark to dare go in Dorothy's room at night . . . and her dread of water was greater than her fear of the dark, so she never came near when Dorothy was bathing.

–Wonderful Wizard of Oz

Region VII of FEMA is headquartered in a burgundy-brick building on Ward Parkway in Kansas City, Missouri. Responsibility for Greensburg falls under this region's purview. The Region VII Administrator is a tall, lanky man named Dick Hainje, a native of South Dakota. Appointed to his position in 2001, Hainje had previously served with Sioux Falls Fire Rescue, a post that he held for 24 years.

FEMA headquarters in Kansas City had been in a conference call with the National Weather Service during the afternoon of May 4. It was obvious that something horrific would happen to anything in the path of a developing tornado. The organization wanted to be aware of anything untoward in the offing. Hainje had talked to General Tod Bunting of the Kansas National Guard early on Saturday morning. He told the General that FEMA was prepared to assist in whatever way it could. FEMA officials began brainstorming Friday morning to try to decide what would need to be done.

It became clear to Hainje that "the only logical thing to do would be to go out directly to Greensburg." He had never seen the town before he began the long drive from Kansas City late on Saturday morning, May 5, 2007. En route, Hainje began to receive phone calls, not only from government officials and his own staff back in Kansas City, but also from the national media. CNN and the Fox Channel wanted to know what FEMA was going to do. Somehow Hainje managed to contact every member of the Kansas Congressional delegation while he drove. He found himself reflecting on some of the previous disasters he had worked on. These included one that had virtually wiped out an entire town in South Dakota when he was with Fire Rescue. Hainje had also worked for four months in Hancock County, Mississippi, in the aftermath of Hurricane Katrina. The phantom of bungled operations north of the Gulf of Mexico in the wake of that hurricane was haunting the staff of FEMA's Region VII.

Dick Hainje arrived in Greensburg in the early evening. There were tornado warnings everywhere. He saw one tornado pass behind him and another one just above US 54, perhaps the one that Jerry Diemart and his nephew, Corey, had outrun in reverse. It had taken Hainje so long to get to Greensburg, what with the stops he had made to place and answer telephone calls, that a couple of his staff were already in town before him. It was getting late. Operations were being shut down for the day. Hainje ran into Angee Morgan, coordinating officer for the State of Kansas. They both made their way to the basement of the courthouse—the only public building of any significance left standing in town—in response to another tornado warning. It was there that Hainje ran into Lonnie McCollum. This is how Hainje recalled the encounter:

> Lonnie and I had quite a conversation standing right there in the basement of the courthouse, lit by emergency lights. The subject of rebuilding green actually came up then. I told him that it was important for a town like Greensburg to have something like a theme in its recovery, something that is unique to that community or to that disaster.

For its part, the National Guard moved in, clamping down on the movement of vehicles and pedestrians throughout town. Apart from officials like McCollum and Hewitt, who were permitted to stay on, making do as best they could, residents were required to leave. Those restrictions were lifted briefly the following Monday when residents were allowed back into town from 8 a.m. to 6 p.m. At first, FEMA had been reluctant to let townsfolk back in so soon after the tornado, but McCollum stood his ground and persuaded FEMA to relent. Registration at the KDOT building, where people had to give an address and their reason for returning, was a preliminary requirement to reenter the town.

Early on, Governor Sebelius criticized the federal government for its deployment of National Guard personnel and equipment to the Middle East. This, she averred, undermined the Guard's ability to respond to the crisis in Greensburg.

"I don't think there is any question if you are missing trucks, Humvees and helicopters that the response is going to be slower," she told CNN. "The real victims here will be the residents of Greensburg because the recovery will be at a slower pace."[41]

White House spokesman, Tony Snow, responded to Sebelius' criticism by saying that "there have been no requests to the National Guard for heavy equipment . . . If you don't request it, you're not going to get it." In fact, however, the State of Kansas had requested an urban search and rescue team, a mobile command center and several helicopters, all of which Snow later acknowledged.

The *New York Times* commented on the spat the day President Bush himself visited Greensburg:

> The debate was reminiscent of the Bush administration's skirmishes with Gov. Kathleen Babineaux Blanco of Louisiana, also a Democrat, after Hurricane Katrina.
>
> . . . . . .
>
> As State Senator Donald Betts Jr., Democrat of Wichita, put it: "We should have had National Guard troops there right after the tornado hit. . . . The response time was too slow, and it's becoming a trend. We saw this after Katrina, and it's like history repeating itself."[42]

In the same article, the *Times* cited evidence which corroborated Betts' complaint:

> For nearly two days after the storm, there was an unmistakable emptiness in Greensburg, a lack of heavy machinery and an army of responders. By Sunday afternoon, more than a day and a half after the tornado, only about half of the Guard troops who would ultimately respond were in place.

By contrast, the paper's remarks about FEMA's response to the disaster were more positive:

---

[41] John Milburn, "Kansas Gov.: Tornado exposed Guard holes," *Associated Press*, May 8, 2007.
[42] Susan Saulny and Jim Rutenberg, "Kansas Tornado Renews Debate on Guard at War," *NYT*, May 9, 2007.

The Federal Emergency Management Agency, which came under strong criticism after Hurricane Katrina, seemed to respond more quickly in Kansas. Several of the agency's mobile disaster recovery centers are in Greensburg assisting residents, and the agency said it had moved in 15,000 gallons of water and 21,000 ready-to-eat meals, enough to feed 10,000 people.[43]

The National Guard's General Bunting said that approximately 850 soldiers had been deployed to Iraq and Afghanistan. The Kansas Guard was equipped to 40 percent of "its necessary levels," down from a pre-war 60 percent. He told reporters that Kansas would ask National Guard units in other states to provide specialized soldiers and equipment to assist with the recovery effort.

"It just leaves you pretty tight," Bunting admitted. "We're fine for now." The *New York Times* quoted the Adjutant General as saying that the Guard had about 350 Humvees—rather than its ordinary complement of 660—and fifteen large trucks instead of the thirty it would normally have had.

****

Tod Bunting looks the way one would expect a major general to look: fit and compact. He and his wife Barbara live south of Topeka on a ranch where they raise driving horses. Bunting comes from a Midwest family—his father from Missouri and his mother from Manhattan, Kansas. He is known as a classic car buff. Bunting became the adjutant general of Kansas at the beginning of 2004. In disaster situations, like Greensburg's in 2007, the State Department of Health and Environment and the Highway Patrol report directly to him.

General Bunting and his Director of Public Affairs, Sharon Watson, arrived by Black Hawk helicopter at 6 a.m. the morning after the tornado. This was the first of several trips which the Black Hawk helicopter made over the next week. Dick Hainje accompanied Tod Bunting on all but the first of them.

By Sunday night, there were two to three dozen Guardsmen in Greensburg, a force that eventually grew to about 600 over the following two months. One hundred airmen from McConnell Air Base in Wichita were in the town as well. The Guard's first priority was security. It was necessary—in military parlance—"to secure the town." Access to Greensburg was strictly restricted over the weekend. In the days that followed, a curfew was imposed. Still, as Bunting said when interviewed at his HQ in Topeka,

---

[43] *Ibid.*

Adjutant General Tod Bunting

"We are sensitive to the need of people to return to their homes." Security and logistics became top priorities: "There was no logistic support within 30 miles of Greensburg," Bunting asserted. Nonetheless his call for a battalion of engineers was heeded within 48 hours.

Sharon Watson returned to Topeka Saturday night, but she drove back the following day, in a van which became her office-on-wheels for the next five days. She was lucky to find a motel room in Pratt which she shared with a Guardswoman. General Bunting himself stayed on in Greensburg most of that first post-tornado week. The Guard, bivouacked on the high school football field, prides itself on its self-sufficiency. Guardsmen bring in their own chow which they serve in their own canteens. They import their own fuel.

It was fiendishly difficult to cope those first few days after the storm. The weather did not cooperate. The storm that hampered the Gardiners' mission to the ruins of the Kiowa County Hospital drove General Bunting and his staff to seek shelter in a FEMA trailer. He recollected that:

> It was like a river coming down the street. Four to six inches of rain which turned to hail. The hailstones hitting the roof of the trailer sounded like a jackhammer. Someone—I can't remember now who—said 'nothing left but the locusts.'

Instead of locusts, people in Kiowa County had another insect to deal with. The *Wichita Eagle* reported one week after the tornado that:

Greensburg residents now have to worry about mosquitoes. The blood-sucking pests are out in full force from the recent rains that drenched the area since the tornado.[44]

At one point the National Guard came across six coffins, scattered higgledy-piggledy along a street. Tod Bunting and a couple of his staff played paper/scissors/rock to choose who would determine whether the coffins were occupied or not. As luck would have it, they were all empty.

Streets were so littered that Bunting's Sergeant Major, Steve Rodina, suffered six flat tires in his first few hours in Greensburg. He told Bunting that he felt like a NASCAR driver.

One day a Guardsman came across a stream of beer cans which spilled over the ground. He followed them back to what had been Charlie and Pat Jones' house. There Charlie was taking his ease under the shell of a silver maple tree. The Guardsman wondered out loud what drunkard could have tossed out so many beer cans. Charlie Jones snorted. These were a few of the many, many Coors cans which he had assiduously collected in the course of his recycling job. Snatched up by the tornado, they had been flung helter-skelter more than half a mile from the Joneses' house.

****

Crystal Payton, an External Affairs Officer with FEMA Region VII, was winding up a couple of months' assignment in Missouri where ice storms had wrought havoc, when the tornado hit Greensburg. She literally did not have time to unpack her bags before being reassigned to the stricken town. Payton arrived in Kiowa County early in the morning of Sunday, May 6, 2007. There she would spend the next month as a Disaster Assistance Employee (DAE). Like Dick Hainje, Payton had worked down in Mississippi after Hurricane Katrina. She had, in fact, logged 15 months there—Crystal Payton referred to herself as a "disaster gypsy"—but her job in Greensburg put her for the first time "in the front lines." The bad weather which bedeviled the early search and rescue efforts of FEMA and other government agencies forced Payton and others into the same courthouse basement where Hainje had met Lonnie McCollum. The ceiling leaked. Everyone in the basement found themselves standing ankle-deep in water. There would be two inches of standing water in that basement for the next two weeks.

Within one week of FEMA's arrival, a bevy of trailers, with their own power generators and computer stations, was clustered around the Kiowa County Courthouse like puppies around a nursing bitch. The biggest of

---

[44]Deb Gruver, from the Greensburg blog, *Wichita Eagle*, May 11, 2007.

these was the FEMA mobile command vehicle nicknamed Red October because it reminded employees of the submarine in the movie by that name. It contained satellite communications equipment and provided IT and logistical support for FEMA. Red October would remain in Greensburg until June 17, 2007.

****

Angee Morgan grew up in north-central Kansas. Like Steve Hewitt, she went to Fort Hays State University, graduating in 1981. She began working soon thereafter for the State Highway Patrol where she met Lonnie McCollum. At that time, he was an aircraft trooper based in Hays. Although their professional paths diverged, Morgan and McCollum remained friends throughout their careers. Years later, McCollum said that he had invited Morgan on what he called a "blood run" from Hays to Kansas City. A 'run' meant transporting a quantity of blood—often a relatively rare kind—from where it was stored to the hospital where it was needed.

KDEM's Angee Morgan

In 1987, Morgan snagged a job with Emergency Management. She was now working in Topeka, in the compound which houses the National Guard, next door, in fact, to the building with Tod Bunting's office. About one year after her relief work in Greensburg, she was promoted to Deputy Director of the Kansas Department of Emergency Management (KDEM), a division of state government which reports to Bunting. A tall blonde woman, Morgan was married in 1988. For their twentieth wedding anniversary, her husband had sent her twenty roses. Four years before she was interviewed for this book, Morgan underwent a double mastectomy. She began chemotherapy the following month. Her recovery was impeded by complications—"I had bad blood counts," she said—which required bone marrow extraction. The pain was so excruciating she could understand why terminal cancer patients chose to discontinue treatment.

Angee Morgan's windowless cubicle in the National Guard compound contained a bookcase crammed with figurines from the *Wizard of Oz*: Dorothy, Toto, the Tin Man and the Cowardly Lion. They were all there.

Morgan drove out to Greensburg on Saturday, May 5, 2007. She was amazed at the extent of the destruction, the worst that she had ever seen. "It was a challenge to get resources because the people I was talking to, at the other end of the line, 100 or 200 miles away, just couldn't conceive of a situation where everything, but everything, was gone."

Morgan's intent, when she went down to Greensburg, was to do damage assessment in order to request federal assistance. She anticipated being on site just a few days but quickly realized that a longer stay would be necessary. Not surprisingly it was hard to find a place to work. At one point, Emergency Management tried to use space on the third floor of the County Courthouse, but the team was advised that this was not safe. Atrocious weather continued to batter the region for days. Over a period of 48 hours, she counted seven times when she had to seek refuge in the courthouse basement. It was while she was working at the courthouse that she encountered her old friend, Lonnie McCollum. His appearance shocked her. "God, you look like hell," Morgan told him. She asked him what he was doing in Greensburg, and McCollum told her that he was mayor of the town.

It rained all morning the day George W. Bush visited Greensburg. Morgan confessed that she wanted to look her best when she met the president, but "that kinda' went by the wayside." She went on to say:

> I'm not a really political person, but to meet the President of the United States was quite an honor. He was very approachable and sincere. I wish I hadn't been so tired because I would have appreciated the moment more. But I kept thinking of all the things I needed to do next.

For her first few weeks in Kiowa County, Morgan lived in a Dodge City motel room. Then FEMA brought in a set of box-like structures called COGIMs. These prefabricated steel-frame modules are manufactured by an Italian company which supplies the United Nations as well as the British,

French, and Italian armies. Morgan moved into her COGIM in early July and loved it. She had a TV set and a DVD player, and she brought from home some of her own pictures and linens, even an easy chair. A photo on the FEMA website, attributed to Crystal Payton, shows the interior of a COGIM much like Morgan's which was only twenty feet from her on-site office in a Highway Patrol command van.

Meals were irregular. In any event, Morgan was not a "breakfast person." She would try to eat dinner at 6 or 7 p.m. Then she could de-stress with colleagues from her own and other relief agencies. "Things that weren't funny at 10 in the morning were hilarious at 7 in the evening," she said.

> Our first priority was to account for all the residents of Greensburg. When the flooding occurred, Greensburg wasn't so wobbly. I guess that we were at the stage of taking off the training wheels.

Shortly after arriving in Greensburg, Morgan attended a Mother's Day service in Davis Park. Her husband and both her daughters, ages 12 and 14, came out to join her. There were hundreds of people in attendance, and the memory of a Mennonite man, singing *a capella*, still had the power to bring tears to her eyes.

Reflecting on her time in Greensburg, Morgan said that she could not separate herself from the townsfolk. They were a special community:

> I've always felt that, even on a daily basis, there are two kinds of people: You're either a victim or you're a survivor, and you choose in the morning what you're going to do. Victims usually lie down and die or they sit around and whine and let someone else take care of them. I admire survivors. I am a survivor, and the people of Greensburg are survivors.

****

Hospital administrator Mary Sweet had been attending a professional meeting in Topeka the day of the tornado. When she learned of the destruction of her hospital, she stopped in Wichita on her way back home to buy $1,000 worth of supplies—water, duct tape, and tarps among other things. Her purchases helped to save 90 percent of the hospital records. The Kiowa County Memorial Hospital had lost more than its physical building; 68 of its employees—including, of course, Chris Gardiner—out of a staff of 95 had lost their homes to the tornado.

The National Guard, stationed at Topeka's Forbes Field, brought in a complex of nine tents called an Expeditionary Medical Support System (EMEDS). These tents—'Alaskan shelters' in the Guards' lingo—were an updated version of the MASH units in Robert Altman's TV series about the Korean War. Deployed on May 11, 2007, a week after the tornado, and operational five days later, the combat hospital received its first patient on May 21, 2007.

It took only four days for civil engineers of the Air National Guard to erect the arches and the purlins for the tents and then to pull the canvas skin over the steel skeleton. As luck would have it, staff from the 190th Refueling Wing at Forbes had undergone training in Michigan to put up these units only one week before the tornado in Greensburg. Still, as Lieutenant Colonel Tim Stevens explained, it was harder to assemble the EMEDS on the terrain in Greensburg than it was in the group's Michigan training program.

The EMEDS was set up on US 54, partly on the concrete pad where previously the Christian Church had stood. The first thing that the Forbes-based Guard had to do was to clear the ground of debris. Stevens said that they were able to salvage some items from the church—hymnals, some serving dishes, a cross, and folding chairs. Then, because the EMEDS was resting on a concrete slab rather than bare ground, holes had to be drilled to secure the arches. Equipment like hospital beds came with the tents. Power was provided by a diesel-operated generator. Tapping into a city water line provided a pump-driven source of water. (The town's water tower, recall, had collapsed in the tornado; to drill into the town's water lines was the only way to obtain water for the facility.) It was somewhat harder to locate Greensburg's sewer line. That operation took almost as long as it did to erect the tents.

There is something called a Quick Set-Up Guide—like a 'how-to' manual—which the Guard follows, step by step, in putting together the interlocking shelters. What it gave Greensburg was called an EMEDS+25 package, so named because the installation can serve 25 surgical and medical inpatients. The first tent to go up housed the emergency room, usually followed by a surgery, but no operations were to be performed at the Greensburg facility. That part of the EMEDS was eliminated. The whole facility still had a price tag in excess of $1 million.

Stevens, a former high school English teacher, has worked full-time for the National Guard since 1983. He had been stationed at Forbes Field since 1987. A native of Chatham, Kansas—a town which itself sustained significant damage from a tornado in 2008—Stevens worked for two years, 2004-2006, in Armenia. His official title was 'Bilateral Affairs Officer', a sort of military equivalent of a health care administrator. He worked in the context of a program which partnered Armenia and Kansas. One goal of the program was to ingrain the American notion that the military is strictly subject to civilian authority. While in Armenia, he met the woman he later married.

"Staffing [at Greensburg] was local, and all we did was to provide logistical support," Major General Bunting said. This contrasted with the situation in New Orleans and vicinity where staffing was provided for a full month after Hurricane Katrina. Most of what was required during the summer of 2007 was treatment for lacerations and the kind of accident which had befallen Megan Gardiner, sustained by people falling down basement steps, trying to salvage what they could from their ruined homes.

****

Senator Pat Roberts, the junior senator from Kansas, was awakened in the wee hours of May 5, 2007, and told of the Greensburg tornado.[45] He lost no time in driving down to the town, arriving at the KDOT refugee center where he encountered an older man dressed in trousers and a T-shirt. "I lost my wife, and now I've lost my house," the man told Roberts. Roberts assured him that "we're going to give you choices." Without hesitation, Roberts drove down the road to Pratt where at the local McDonald's, he placed a phone call to the White House. Roberts insisted on speaking directly to the President. When George W. Bush came on the line, Roberts said, "Mr. President, *we can't have another Katrina*," to which President Bush replied, "The green light's on."

The President came to Greensburg on Wednesday, May 9, four days after the tornado. He arrived by helicopter shortly after 10 a.m., then made his way by motorcade to the John Deere dealership where he mingled with townsfolk amidst the twisted hulks of combines and tractors. From there, he continued to the Kiowa County Memorial Hospital. Next to the ruins of the United Methodist Church several clergy joined the president in prayer. One Greensburg resident, Vernon Davis, who had come to see the president, was surprised to be approached by a female FBI agent. "The president wants to talk to you," she told him. When Davis walked over to Bush, the president asked, "You're a farmer, aren't you?" Davis said that he was. He did dryland farming. So no corn, just milo and wheat. "How did anyone survive here?" Bush asked him. "We had 20 minutes warning," Davis replied. Then he added, "God has given man a lifetime to prepare for that twinkling of an eye." Bush wanted to know what church Davis went to. Vernon said, "Lighthouse Worship."

---

[45] Roberts is regarded as the dean of the Kansas congressional delegation. That he was the junior senator from the state is something of a technicality: His colleague, Sam Brownback, was sworn in as Bob Dole's replacement in 1996, on the night of the election when Roberts first won his Senate seat. Upon Brownback's retirement to run for the governorship of Kansas, Roberts became the state's senior senator.

Lonnie McCollum accompanied the president from the airport where his helicopter had landed to the John Deere dealership. "He's a genuine man," McCollum said, "I think he really does care about people here."

The president summed up his mission during his visit as one "to lift people's spirits . . . and to let people know that while there's a dark day in the past, there's brighter days ahead."

As a souvenir of the presidential visit, Vernon Davis received a photo of the president's helicopter hovering over Greensburg.

Near the end of his visit, President Bush met with the national director of FEMA, R. David Paulison, who arrived in a Wichita city bus, accompanied by officials from that city. This was the meeting for which Angee Morgan had hoped to look her best. It took place in Red October and included meteorologist Mike Umsheid, Governor Kathleen Sebelius, and Dick Hainje.

****

By the end of May, 230,655 cubic yards of debris had been hauled away to nearby land fills, enough to have filled 64 Olympic-sized swimming pools. By the time the dust—or in the case of Greensburg, the mud—had settled on the clean-up, 600,000 cubic yards of debris were removed from Greensburg. The effort, according to FEMA's Crystal Payton, required 1200 truckloads of debris removal a day.

High school graduation was rescheduled for Saturday, May 19, 2007. Since the High School itself lay in ruins, the event was held on the eastern edge of town, in a complex of tents erected beside the Greensburg golf course. Commencement speaker Senator Pat Roberts told the 25 graduating seniors that they were "a class of destiny and hope." He observed that this Saturday "was a day you will remember for the rest of your lives, a day unlike [that of] any other graduating class." He referred to May 4 as a day which had put the seniors "through a storm and an ordeal unique to the history of Kansas. . .What we have here . . . is a whole lot of people who specialize in the impossible." He offered individualized bits of advice to graduates toward the end of his speech, cautioning one young man not to drive his car in the lake and another to continue his singing career by choosing 'Home on the Range' over 'Oklahoma.'

Another commencement speaker was US Representative Jerry Moran. Greensburg lies in his Congressional district. Noting that he had been the commencement speaker in 1997, Moran said, "It must not have been very good ten years ago because I haven't been asked back until now."

Moran paid tribute to Mike Umscheid who had decided to "push the button" and issue the tornado emergency for Greensburg the night of May 4. "He was at his [own] high school graduation just a few years ago," said Moran, "and I'm sure he never dreamed back then he'd soon be making a decision that would save hundreds of lives."

High School Principal Randy Fulton told the seniors that the awards which some of them were to have received "were typed up on my desk three weeks ago, but they're now gone. I'll have to get new ones to you guys later."

All the school district's facilities had been levelled by the tornado. In fact, the only school property to escape destruction were a Chevy Suburban and a motor coach which had been out of town on Friday, May 4, 2007. Yet a commitment to reopen in time for the following school year was kept. Superintendent Headrick told the Topeka *Capital-Journal* that 28 modular classrooms and four modular offices were in place by August 12, 2007, the time when school reopened, with 74 percent of its student body in attendance.

****

His name was Bond. James Bond. Not surprisingly, he took some ribbing for that moniker. A contractor by trade, Bond hailed from Manhattan, Kansas, but he had been born in St. John, only one hour northeast of Greensburg. He describes himself as a Christian man. "God has given me the ability to build and to teach," he said. With what he described as his family's support—Bond was married and had three daughters—he shuttered his business in Manhattan in order to assume the post of project manager for the South Central Kansas Tornado Recovery Organization. Its acronym, SCKTRO, is no easier to say than the name of the organization itself. Commonly called South Central, it came into existence in response to the clean-up in Kiowa County and four nearby counties. It is an almost exclusively faith-based consortium of relief groups although there was participation by United Way of the Plains and the Cannonball Trail American Red Cross. It was not uncommon in situations like Kiowa County's to create a relief task force under the auspices of an existing organization. SCKTRO was birthed under the patronage of the Greensburg Ministerial Alliance. That way, the organization could benefit, without dealing with masses of red tape, from the Alliance's tax-exempt status.

James Bond came on board SCKTRO in early July 2007, less than two months after the tornado ripped Greensburg apart. His title was that of general construction manager. He stayed in that position for less than one year, leaving at the end of March 2008.

Bond's job was to coordinate the work of volunteers who poured in from all parts of the country to help build houses for the homeless residents of Greensburg. Volunteers were housed in trailers and fed by organizations like the Salvation Amy, the Red Cross and the Southern Baptists. One week in March 2008, there were 400 volunteers working under Bond's supervision. The week before, it had been 200. Another 200 came the week following. James Bond dealt with the faith-based groups, like a contingent from the Wesleyan Church in Sacramento, California. That particular group was hammering and nailing away on a house for Charlie and Pat Jones, north of US 54, when Bond was interviewed for this book. The same day he showed some students how to run an earth compactifier at another job site half a mile away. That demonstration generated equal measures of giggles and *angst* among the volunteers.

Volunteers working inside the Joneses house

The contractor's life as project manager of SCKTRO was far from smooth sailing. At a meeting of the City Council in September 2007, less than three months after his arrival, Bond came in for some harsh words from a resident frustrated at the delay in assigning him a case worker. Bond cited the sheer number of requests to flood SCKTRO, then juggling with more than 100 cases. He promised the man that his volunteers would get to him as soon as possible. At the same time, Bond was scolded for the lack of communication between his organization and the city. What brought about the rebuke was an incident in which a number of SUVs belonging to SCKTRO volunteers had been ticketed for parking in Davis Park.

"Your group has been off the radar screen for quite some time," John Janssen, the man who succeeded Lonnie McCollum as mayor of Greensburg, told Bond. "We need a memorandum of understanding between you and us so we can keep up on what your group is doing."

Janssen had been critical of South Central from the get-go. At one interview in 2008, he said:

> There was no communication with the community, and they were doing things . . . they didn't have the resources or the experience or the wherewithal to do. They actually kept Habitat for Humanity out, and it took a while to get that all straightened out.[46] South Central was created by Ministerial Alliance to counsel people who needed food and shelter. There was a blank that needed to be filled and it would have been good if South Central could have filled it. But they decided that that wasn't glamorous enough. If volunteers came into town and I had anything to do with them, I ran them through Mennonite Disaster Services [MDS] because MDS supported them and made sure that they had a job to do.

Many of Bond's volunteers had had no previous experience in construction work. For them as well as for the others, Bond's goal was the same. "My hope is that they will take away the feeling that they're helping others," he said. "That's what makes this country different. . .when we get it right."

****

The Mennonite Disaster Service (MDS) to which John Janssen referred volunteers is an organization with a long history of relief assistance. Effectively the MDS was a competitor of the Ministerial Alliance's SCKTRO. Local director for the MDS since 2005 had been Kenny Unruh. Months after the tornado, he talked about the organization and what it had done in Greensburg.[47]

MDS began at a 1950 Mennonite picnic in Hesston, Kansas. It has binational offices in Pennsylvania and Manitoba; a modest budget—$20 million in disaster relief for the victims of Hurricane Katrina; a small staff—only

---

[46] In an exchange with John Janssen at a City Council meeting of September 20, 2007, Bond said that Habitat had not been "very aggressive in getting involved." Janssen then went on to say that Habitat had been told to go away. "That didn't come from me," Bond shot back. Janssen later told the *Signal* that his source of information had been a Habitat representative.

[47] Kenny Unruh is the son-in-law of Gordon Unruh.

thirteen in Pennsylvania and three in Manitoba; and two field consultants—Doreen and Jerry Klassen. Mennonite churches budget for disaster relief. Some companies donate materials. One in Michigan, for example, sent a 45-foot reefer to Greensburg filled with everything from 2x4s to apples. "We went around asking for apple recipes," Unruh said. He is himself a member of the Mennonite Church in Greensburg, an offshoot of Bethel, the Holdeman congregation south of town. An engineer by profession, Unruh volunteered to head up the relief effort in Greensburg the Sunday following the tornado.

Mennonites are well known for their thrift. They throw nothing away if they don't have to. "Copper wire was invented by two Mennonites out of a penny," Kenny Unruh said with a gut-splitting laugh. Mennonite volunteers were reluctant to discard what they found in the debris of Greensburg. One young man couldn't bring himself to toss away a brand new pair of tennis shoes. Although Unruh had cautioned his volunteers not to take anything without the owner's permission—and not to ask for anything, either—he consented to speak to the owner of the tennis shoes. The problem, Kenny discovered, was that the shoes were studded with glass shards.

MDS volunteers were sometimes lodged with the Unruhs whose house, at one time, held 35 people. The house where the interview with Kenny took place cost the MDS originally $5,000 with an additional $20,000 for materials to restore it. To apply for a house, residents needed to show that 1) they had owned their own place prior to the tornado, and 2) they were currently unable to afford a new house. The MDS helped individuals this way. Not so landlords with multiple properties. There was one instance of a man with 50-odd properties around Greensburg who had requested assistance and was turned down. Although 65 percent of the Mennonites in Greensburg lost their homes to the tornado, the MDS built only one house for a member family of the congregation. To finance its reconstruction work, the Mennonites wrote to the Greensburg Future Fund (GFF), an organization set up under the auspices of a group called the South Central Community Foundation. Its Executive Director was a local woman, Denise Unruh. The MDS requested $65,000 to help low-income families who fell through the cracks at SCKTRO. The grant was approved and funded as one of several projects supported by GFF, whose total cost exceeded $500,000 (a figure valid as of April 15, 2008).

Relations with SCKTRO were sometimes strained. Unruh found James Bond difficult to get along with. "He thought that he was the savior of Greenburg." At first blush, it would seem that the combination of SCK-TRO's money and MDS volunteers was a marriage made in heaven, but conflict developed over lines of authority. To facilitate relations with South Central (as SCKTRO was commonly, if confusingly, known), Unruh attended a meeting in April 2008 with James Bond. Somehow the meeting expanded into a confab which included a representative of USDA. Harsh words were exchanged. "You don't know what you're doing," Unruh was told by the government rep. This was followed by "you have $100,000 in government money to spend, and nothing has so far been spent."

Kenny Unruh is a man with a sense of humor. He recounted anecdotes which gave a bit of the flavor of the weeks immediately following the tornado. One day the phone at the MDS trailer rang. "Unruh," one of Kenny's friends yelled, "someone on the phone wants to know if there is anyone here who speaks Mennonite." How this request got to Greensburg is a story in itself: The call was placed by a Florida man working for a company which had a translation contract with a hospital in Dallas, Texas. A young Mennonite woman at the hospital needed emergency surgery. It was her mother who had contacted the company in Florida. The patient needed to be told what sort of surgery was required. English was not a *lingua franca* since both she and her mother came from Mexico where there is a well-established Mennonite community south of Ciudad Juarez. Kenny's rusty dialect of Platt Deutsch was not up to the task. His rendering of *Kanst Du Deutsch verstehen?* was greeted with thunderous silence. Fortunately there was at hand a Canadian volunteer who had lived in Mexico. His version of Platt Deutsch, supplemented by Spanish, did the trick. How had the Florida company obtained the telephone number of the Greensburg Mennonites? Apparently it was the first to pop up in the MDS directory. The Canadian volunteer received $250 for his translation services.

Volunteers came to the MDS from all faiths, not just Mennonite, although religious groups theologically close to the Mennonites were the most prominent in the Greensburg clean-up drive. One group of Amish men came from Missouri. They had left their homes at 2 a.m. to arrive four and a half hours later at the check point set up by the National Guard east of Greensburg. Thirty-four volunteers showed up this way, each toting a suitcase, containing no more than a change of clothes and hand tools. One man came with a six-foot spud bar. Some of the men were illiterate. They couldn't sign their names in the roster of volunteers kept at the check point. But they were efficient workers. Kenny Unruh divided the men into two groups, each under the supervision of an Amish bishop. The first group was set the task of tearing down a building and clearing out its basement. While men pushed on a wall, an Amish kid grabbed a Romex cable and used it as a tether to help bring down the wall. Eight men carried the 35-foot remnant of wall to the street. "Should we have done that?" the bishop-in-charge asked Kenny anxiously. "Do you have machinery to handle such a big piece?" The second group was bussed to a site where Kenny Unruh observed three men up on the rafters of a building to be demolished—wobbly rafters at that—before the last man got off the bus. No ladder was used. "We are an agile people," explained the bishop-in-charge as a matter of fact.

Tales of the Amish work ethic were legendary. Weeks after the tornado, a group of Amish men was asked to salvage the wood of an 8-foot by 30-foot deck from a widow's house. The woman's brother-in-law wanted to rebuild the deck at his place. The Amish salvaged the wood—and how: They were able to save the deck without dismantling it. "It wasn't easy to save the whole deck," the presiding bishop told Unruh. "It took 20 men to put it on a truck bed."

****

In the early days of recovery, day-to-day operations in Greensburg were coordinated by a group called the Incident Management Team (IMT). Besides city officials, the IMT brought together organizations like the Red Cross and the Salvation Army. Each subdivision of the IMT—administration, finance, logistics, and operations—corresponded to a counterpart at FEMA.

At each meeting of the IMT, the Incident Commander—this was the name given the chair—established the goals to be achieved over the next operational period. This was generally the next 24 hours although it could stretch as far as five days ahead. It was the role of FEMA, described by Crystal Payton as "the bearer of technical resources," to make up for the shortcomings of the IMT. Radio units, for example, brought in by the state to broadcast messages to Kiowa County residents were supplemented by FEMA.

FEMA invited the US Forest Service to set up a base camp in Greensburg. The camp operated for about one month, from May 8, 2007 to June 11, 2007. During this time, it served more than 36,000 meals. Said Dick Hainje:

> When I was flying down to Greensburg with General Bunting, I would always bring along a golf cart, go over to the base camp, and grab a bunch of sack lunches. Sometimes people got more than one meal from different sources, but food is what they needed and wanted. A disaster works on its stomach, after all.

Hainje recalls that he was in Greensburg when the Qwik Shop reopened. He told his travelling companions—who included most of the state's Congressional delegation—that he hadn't had a chance to spend any money in Greensburg yet. So he bought a Slurpee and a diet Coke at the Qwik Shop.

Inevitably there was red tape to contend with. Applicants for disaster assistance needed to acquire first a FEMA registration number, even if they were applying to other agencies for help. If applicants said that they needed housing assistance, inspectors had to ascertain the facts of the matter. In the case of Greensburg, this was little more than a formality, given the damage that the tornado had caused.

The most visible sign of FEMA assistance were the rows of FEMA mobile homes lined up southeast of town. It was at first unclear whether the mobile homes would be put on people's property or clustered at a single location. Both these options were adopted although the bulk of temporary housing was put in the location leased by the agency. This mobile home park, which was formally known as the "Keller Estates," acquired the nickname—what else?—FEMAville with enough space for 500 units, although only 300-odd families applied to live there.

Mobile homes in FEMAville

Dick Hainje said that utilities—hook-ups to water, sewage, and electricity—had to be available before the mobile homes could be moved into place. In the case of Greensburg, this also involved tornado shelters for each block in FEMAville. To save time, FEMA uses a pre-bid awarding of contracts to prepare sites. Construction was still hampered by the exceptionally bad weather which the region experienced through mid-summer 2007. Contractors were constantly up to their shins in mud.

The mobile homes themselves began to arrive in Greensburg on July 12, 2007, a little more than two months after the tornado. Residents could stay in their units for as long as 18 months. After that, FEMA reserved the right to charge rent. When he was interviewed for this book, at the beginning of October, 2008, Dick Hainje said that about 75 units were still occupied. He also repeated FEMA's policy of charging rent 18 months after first occupancy. A few days later the *Lawrence Journal-World* reported that this would, in fact, happen. Rent would be determined by a family's net income, with a minimum of $50 and a maximum of $667 being charged per month. Steve Hewitt was successful in persuading FEMA to alter its rent scale. "I just don't want to lose citizens," he said. What Hewitt accomplished was to get FEMA to charge rent on an incremental basis: $50 for the first month of occupancy after 18 months, $100 for the second month, $150 for the third month and so on until a cap of $650 was reached.

The Gardiners moved into their FEMA mobile home at the end of July. One last quote from Chris Gardiner's tornado poem:

> In July we sipped on a Zima
> Then we got a call from FEMA.
>
> They said our trailer was ready,
> But some of the rules seemed petty.

Although the Gardiners were grateful for the shelter they provided, life in a FEMA trailer left something to be desired. Chris suffered from cabin fever. He would pace from one end of the trailer to the other. He chafed under restrictions like the one which prohibited hammering nails into the walls. In an interview with Chris which the *Wichita Eagle* ran on the anniversary of the tornado, the paper reported that:

> Outside the small [FEMA] trailers were reminders of what they'd endured, and might endure again: Storm shelters, crude steel boxes buried in earthen berms. To the west and north, Gardiner and his daughter could see the shattered tree limbs that survived, twisted and bare of bark, as though sandblasted by the gods.

There were petty nuisances. Each time Julie cooked, the smoke detector over the stove went off. This was a complaint voiced by other FEMA trailer-dwellers. The complaints—and, of course, there were some—were relatively minor. Lloyd Goossen had some amusing comments to make about the trailer which FEMA provided him and his wife: The exterior skirting was not designed to withstand Kansas winds, for one thing, "but it was good to have a roof over our heads." On balance, FEMA got high marks from residents for its assistance program. Roxana Hegeman of the *Associated Press* wrote on September 7, 2007:

Around Greensburg at least, FEMA isn't a dirty word. Four months after a twister practically obliterated the town, . . . many folks have nothing but good things to say about the federal agency that was lambasted as spectacularly inept and clueless following Hurricane Katrina in New Orleans.[48]

Dick Hainje worked for about four months with the agency on the Gulf Coast after Katrina had roared through the region. He described his work as that of a "trouble shooter," liaising with state and local officials and briefing Secretary of Homeland Security, Michael Chernoff, on the situation as it evolved. Hainje brought back with him lessons learned from his experience in the Gulf. Setting up a facility like EMEDS was an example of how that kind of experience came into play: Erecting these MASH-like units had been done in Hancock County where Hainje worked. He said:

I was prepared when I went to Greensburg for there to be absolutely nothing to eat, no place to get gas, no place to sleep, no place to pick up fresh batteries for a flashlight.[49]

Toward the end of his interview, Hainje mentioned, almost as an afterthought, that the agency was working to establish partnerships between US and Chinese cities which faced similar challenges. Greensburg, he said, might be involved in just such a partnership. And that is exactly what happened.

On October 27, 2007, Greensburg participated in *Make a Difference Day*, an event sponsored by *USA Weekend*. It attracted entries world-wide. Two Greensburg women, Ann Dixson and Ruth Ann Wedel, marshalled the town's effort. It brought in 260 volunteers, including 95 town residents and 50 Mennonite farmers who came with badly needed heavy equipment. The effort was recognized by the magazine which made one of its ten awards to Greensburg, giving the town top billing and a cover photo in its April 25-27, 2008 issue.

---

[48] Roxana Hegeman, "Town smashed by tornado praises FEMA's quick help," *Associated Press*, Sept. 7, 2007.

[49] The interview with Dick Hainje took place at FEMA Region VII headquarters in Kansas City, MO, on October 2, 2008.

GOING GREEN

"...since you demand to see the Great Oz I must take you to his palace. But first you must put on the spectacles."

"Why?" asked Dorothy.

"Because if you did not wear spectacles the brightness and glory of the Emerald City would blind you."

–Wonderful Wizard of Oz

It was not long before meetings to discuss the future of Greensburg were called. In an interview in *The Signal* on May 9, 2007, Mayor Lonnie McCollum chose to put a brave face on things. "We have the opportunity for a brand new town," he said. Photographed wearing a white T-shirt and his signature baseball cap with its panoply of American-flag stars, McCollum acknowledged that the town was not going to come back overnight. "But in five years, there will be a huge improvement."

On Friday, May 11, 2007, exactly one week after the tornado, there was a town meeting in Davis Park. Five hundred residents of Greensburg attended. According to *The Signal*, it "could have passed for a frontier revival meeting in its opening minutes." Neighbors who had not seen one another for a week fell into each other's arms and wept for joy. The crowd heard McCollum and Hewitt offer words of encouragement to those whose lives had been shattered by the tornado. Volunteers were thunderously applauded. The City Administrator paid tribute to McCollum. "The mayor is going to start construction next week on his brand new energy efficient house in Greensburg." The announcement brought the crowd to its feet.

"We're going to rebuild this town, and we're going to do it right," said McCollum.[50]

Darin Headrick assured people that the school year would be brought to a successful conclusion. The new school year would begin as scheduled in mid-August. Hospital Administrator Mary Sweet was reported to be working to get a portable hospital operating by the end of the week.[51]

The rip tides underlying all the optimism at the May 11 meeting were a more somber matter. The tornado and its immediate aftermath took its toll on Lonnie McCollum. His wife, Terri, had hurt her knee so badly that it required corrective surgery. His house, now in ruins, had been woefully underinsured: $175,000 for property which Lonnie valued at $700,000. Friends actually shipped McCollum off to the resort of Branson, Missouri, for three days of R&R. That may have helped but it did not suffice. A city council meeting on May 21 proved to be the straw that broke the camel's back.

At first blush, the specific issues raised at that meeting seemed unrelated to the question of going green. The first of these was suspending the neighborhood revitalization act. The second was the imposition of a graduated building permit fee schedule. In fact, Lonnie McCollum, after consultation with Steve Hewitt, had decided to rescind both these actions before the meeting of May 21, but that did not defuse the resentment that was ready to boil over. Leading the opposition was a businessman called Kelly Estes. In partnership with his brother Mike, Estes was co-owner of the John Deere dealership, known locally as BTI John Deere. The initials stand for Bucklin Tractor Implements. Bucklin is the name of the nearby town where the Estes brothers' grandfather had established the company in 1944. Since that time, BTI John Deere had remained in the hands of the Estes family, fathers passing on the reins to sons, for three generations of the family with the fourth generation waiting in the wings to take over.

Kelly Estes presented each Council member at the meeting of May 21, 2007, with an inch-thick sheaf of documents comparing the actions of the Greensburg City Council with positions adopted by other communities.

Estes cited the two issues mentioned above as well as the expense of an across-the-board building code. Halfway through this harangue, McCollum said that he had already decided to veto the objectionable actions, but—in the mayor's words—Estes "just kept right on going." The following Friday, May 25, McCollum called his friend, Mark Anderson, at the *Signal* to

---

[50]The Nov. 5, 2009, issue of *The New York Review of Books* contains a review by Bill McKibben of Rebecca Solnit's book, *A Paradise Built in Hell: The Extraordinary Communities that Arise in Disaster*. The author recounts instances of selfless generosity which followed the San Francisco earthquake and fire in April 1906.

[51]This was the EMEDS set up by the National Guard.

announce his resignation. He had earlier told fellow Rat Packer, Gary Goodman, of his decision, and Goodman spent several hours with the mayor "driving around the state of Kansas," trying to persuade him to change his mind.

McCollum later described the May 21 meeting as "bickering." He was despondent about the future of Greensburg's going green:

> Now the momentum in town is for letting everyone put it back the way it was, and now they think they can if they raise enough hell. . . . The shortsightedness of those who want to do [the rebuilding] fast, cheap and easy is too strong. [52]

On June 6, 2007, McCollum resigned and Vice Mayor John Janssen took his place as mayor. Steve Hewitt expressed regret over McCollum's decision:

> I'm disappointed he didn't hang on a little longer, but he had to do what's right for him and his family.[53]

McCollum might be gone, but the Rat Pack was intact and in control. Hewitt, for one, would soldier on. And McCollum's successor, John Janssen, was known to share the ex-mayor's views about sustainable growth. A Falstaffian figure of a man, Janssen was described as an "anti-politician." He did not mince his words. His tenure as mayor, which lasted one year, was a time in which direction-setting decisions were made about the future of Greensburg.

Janssen was not without support in continuing McCollum's vision for Greensburg. At the town meeting of May 11, a man called Daniel Wallach had circulated what he called a concept paper among his neighbors. Many of them shared his ideas about sustainable living. This included, of course, Lonnie McCollum, Steve Hewitt, and soon-to-be mayor, John Janssen.

> What if Greensburg were. . . rebuilt as a model green community?

That's how Wallach began the first draft of his concept paper. He went on to add:

> This town would be a demonstration center for clean and modern energy, and state-of-the-art resource usage and conservation.

---

[52] Mark Anderson, "McCollum steps down from mayor's post," *KCS*, May 31, 2007.
[53] *Ibid.*

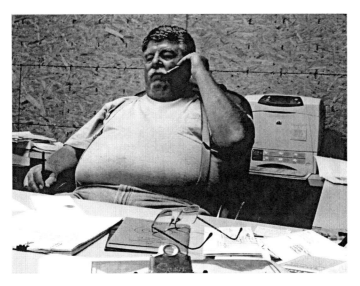

John Janssen

Daniel Wallach became an advocate for the green movement. He was probably the most widely quoted—just as Steve Hewitt became the most visible—of the people involved in that movement. Health problems had compelled Wallach, in 2003, "to simplify his life." He had fallen ill with complications from a childhood bout with encephalitis. What ensued was a 10-year battle with illness, consultations with scores of neurosurgeons and alternative healers. His wife, Catherine Hart, was herself suffering from a brain tumor. Originally from Colorado, they settled in Kansas on ten acres of wooded land in Stafford County, 35 miles north of Greensburg.

Talking to John Janssen at the May 11, 2007, town meeting, Wallach learned that Lonnie McCollum had already introduced the idea of going green. Circulating copies of his concept paper, Wallach encountered both grass-roots support for his ideas as well as a measure of skepticism. With like-minded neighbors, he founded a group called Greensburg GreenTown in the summer of 2007. The group signed a formal memorandum of understanding with the city of Greensburg on August 6 of that year. The activities of this organization ran in counterpoint to what the city of Greensburg did over the following four years.

****

The dominance of the Rat Pack, reinforced by Daniel Wallach, is the reason Greensburg went green. But that is only part of the story. What muscled the green dynamic was a one-two punch. First, the Rat Pack's producers at Planet Green. Be that as it may, people at the Thursday meeting, townsfolk and Council members alike, listened intently to what their visitors had to say. Steve Hewitt and Mayor John Janssen had met with these men hours before. The visitors' comments reinforced one of Janssen's contentions that the reconstruction of Greensburg—Main Street in particular—would need to be eye-grabbing, to set Greensburg apart from other towns in the Great Plains. But Thursday night's speakers offered an even loftier goal: The need to grab the attention of Wall Street investors. Said one Los Angeles-based architect:

> You could have rows of beautiful white wind turbines down the middle of Main Street. Do something unique so big companies can see it and be encouraged to invest in it. The wind destroyed you, so why not now turn the wind to your advantage?[57]

An Atlanta-based financier said: "Wall Street investors are looking at you. Sell your story to the markets so people will come and invest."

With his shock of silver hair and electric blue eyes, Picard held the floor for much of the meeting's three hours. Echoing the previous remarks of the Atlanta financier, he told the crowd that the reconstruction of Greensburg had the potential to inspire a nation desperately in need of a spiritual rebirth:

> To take Main Street and to do something phenomenal with it will attract investors. If you build back conventional, it'll be just that. Building to code is just one step above breaking the law. [58]

John Janssen was quick to jump on the bandwagon that Picard was driving through the Greensburg wasteland:

> We were a dying community. Rebuild [Greensburg] like it was and you doom it to failure. The 'wow' factor is what will sell this town.

Greensburg auctioneer Scott Brown paid homage to the eloquence of the visitors who came to town for the meeting on September 13:

---

[57] Mark Anderson, "Big city planners find favor with many residents," *KCS*, Sept. 14, 2007.
[58] *Ibid.*

People have said I've got a silver tongue, but I don't hold a candle to these guys. The ideas I've heard here tonight are wonderful, but we need people to do this.

The matter of delays weighed on some residents. The business community, Scott Brown noted ruefully, had been without income for four months. "It'd still be 18 months before they'd start to make [money] if they started building tomorrow morning."

For his part, Chris Gardiner—who attended most of these meetings—thought that the presentation of September 13 was just another of the "pep rallies" which the City Council was staging. Certainly that was one of the purposes of the meetings. Decisions made at these meetings, however, affected development in Greensburg. Picard was aggressive in advocating more planning despite the FEMA studies for the town, one of which looked like a master plan to people like Scott Brown.

"We felt like the comprehensive plan we worked on for three months with FEMA was a master plan, and now we hear we need another one," he complained. Steve Hewitt demurred: "A true master plan can help us move forward. . .rather than just continuing to talk." Hewitt went on to recommend that "Greensburg [needs to] look at what the architects can put forward, and then hire someone to put a plan together and have them do it quickly."

The *Signal* concluded its coverage of the September 13 meeting with the following exchange, beginning with a Picard quote:

> "You have to want a town that's special, and we're just here to tell you what's possible. We should design this town for generations not yet born. We get your message, and it's one of urgency."
>
> Picard later looked at Hewitt . . . and told the crowd, "You've got a good guy here in Steve. Steve, you can do this!"
>
> "You're hired!" Hewitt cracked with a broad grin.

The previous month, a call had gone out to firms to submit ideas for city planning in Greensburg. There was no formal request for proposals (RFP), but two dozen people showed up—potential donors as well as architects—to make presentations. They came from all over the country.

One of the companies to make a presentation was BNIM, a firm in Kansas City. This firm had clear advantages over its competitors. Besides its relative closeness to Greensburg, its track record of designing environmentally sensitive buildings included two well-received projects for the

University of Houston: an Institute of Molecular Medicine and the School of Nursing. About 20 years after its founding in 1971, the company, then known as Patty Berkebile Nelson Love Architects, began moving in the direction of sustainable building. At the invitation of Governor Sebelius, BNIM had been working in Greensburg since late May 2007, first with the State of Kansas, then with FEMA.

Not least among the firm's advantages was the long-term relationship between John Picard and Bob Berkebile, a founder and partner in BNIM.[59] (The 'B' in BNIM stands for Berkebile.) At Picard's urging, the Greensburg City Council asked BNIM to draw up the city's master plan on October 7, 2007. In fact, the selection of BNIM was reported to be the *sina qua non* for Picard's continued participation in Greensburg planning. Early in the game, Picard had called Hewitt to say that he simply had to be part of the project of rebuilding Greensburg, to which Hewitt replied "You cost $10,000/day." With a shrug of his shoulders, Picard told him not to worry about payment.

Point man for the BNIM master plan was Stephen Hardy. A 30-year-old urban designer, Hardy joined BNIM after attending Harvard's Graduate School of Design, 2003-2005. Previously he had gone to the University of Kansas. He was, in fact, the fifth generation of his family to attend the state university. He had also worked for Congressman Jerry Moran, whose district included Greensburg, then for the Conservation Fund where he had done environmental planning. Like John Picard, Steve Hardy was to figure prominently in the Planet Green series on Greensburg, particularly in the early episodes.

The August call for city planning proposals "overwhelmed and disappointed" Hardy, who had been engaged with FEMA to produce the agency's own plan which was released in October 2007. Entitled *Long-Term Community Recovery* (LTCR) *Plan*, it reflected views expressed at a series of community meetings; a Design Workshop held on July 17-18, 2007; and a Rebuilding Fair held immediately afterward.

---

[59] Bob Berkebile was thrust into national prominence as a result of a tragedy which occurred on July 17, 1981. Skywalks in the Kansas City Hyatt Regency Hotel collapsed, killing 114 people. Berkebile had been one of the two principal architects for the building. The National Bureau of Standards exonerated his work in determining that engineers, not the design architects, had been at fault. Be that as it may, Berkebile felt compelled to examine the environmental ethics of architecture. Notwithstanding what one Lawrence-based architect described as "an epiphany" for Berkebile, the notoriety which his firm incurred as a result of the accident took its toll. Berkebile's long-standing friend and partner, Tom Nelson, whose last name contributes the 'N' to 'BNIM', was convinced that no company with any connection to the disaster would survive. The firm morphed into BNIM, the name adopted for the company in 1991. For a more complete discussion of the man and his company, see Carolyn Szczekpanski's "Hyatt Regency skywalks designer Bob Berkebile is the godfather of green building," *Pitch*, April 8, 2008. See also the coverage of the Hyatt Regency disaster by the *Kansas City Star*, which won a Pulitzer Prize for local news reporting in 1982.

"This LTCR Plan is a guide for Grensburg and Kiowa County to use in their recovery efforts following the May 4, 2007, tornado," the report said. "The Recovery Plan should be viewed as a guide, not specific instructions." Nonetheless, the plan set the tone for much that happened over the next 18 months. It dwelt, for example, on the importance of alternative energy sources: "Greensburg is located in the High Plains of central-western Kansas where both wind and solar energy stand out as potentially viable sources of renewable energy."

The plan raised what was to be a crucial issue for the town: aiming for LEED platinum certification. It stated that "designing and constructing public facilities to meet the most stringent environmental and energy efficient standards will increase the sustainability and add unique elements to Greensburg and Kiowa County."

Running in parallel to Daniel Wallach's own thinking, the plan observed that "a concentration of LEED platinum buildings in one community could provide a tourist attraction, especially for those interested in. . . the rebuilding process."

The plan also supported the rebuilding of the Twilight Theatre which "would be a key project toward the restoration of Main Street as a vibrant community center and could serve as a catalyst to commercial development in the county." Ironically, in the light of what actually happened, there was only tepid support for an arts center which, according to the Plan, "has a low recovery value and is a complimentary (*sic*) cultural addition to the community facilities centered around the Park Square." The cost estimate for such a center was put at $450,000.

How much of an overlap was there between the two plans drawn up by BNIM and FEMA? "Discussions between townspeople and FEMA took place in midday," said John Janssen. "So only retirees and the unemployed could attend." The most active segment of the population was thus excluded from the decision-making process. Janssen was also dissatisfied with the work done by architects whom Kathleen Sebelius had recruited to help Greensburg.

"I told them to take their charette, roll some marijuana in it, and smoke it," Janssen said with a guffaw.

Even the work done by BNIM was too conservative to suit Janssen. The Planet Green series captures several exchanges between him and Steve Hardy, some genial, others less so. One of Janssen's favorite hobby horses was the inclusion of substantial park space at the intersection of Main Street and US 54. Many merchants opposed the idea. BNIM chose not to make it part of its own plan for downtown development.

The master plan that BNIM offered the City Council was presented in two parts. The first came on Tuesday, January 16, 2008. It was approved the following week. Known as City Ordinance 946, the plan was adopted unanimously. A payment of $83,560 to BNIM and McCluggage Van Sickle and Perry (MVP), a collaborating firm from Wichita, was approved.[60] The payment was made for consulting fees incurred by the city for the architects' work to incorporate a LEED platinum certification into the master plan.

The first part of the master plan "emerged directly from the community and is representative of both the planning team's recommendations and the input from many stakeholder groups." It dwelt at some length on questions of energy use and generation, citing the need to reduce carbon emissions and to promote energy independence as "arguably the most important issues facing our society." Wind energy was specifically mentioned as one way to power Greensburg 100 percent from renewable sources.

The plan called for a "walkable community founded around a civic core," with guidelines for "sustainable site and building design." Buildings should be "healthy, efficient, safe, beautiful structures" which should last for 100 years. Architectural design was to be sensitive to regional influences and to use local materials.

The third issue to be addressed in the plan was the need to treat water as a precious resource. It was observed that the Ogallala Aquifer was being exhausted "at a rate ten times the rate of natural recharge."

> The city will implement sustainable infrastructure solutions for both storm water and waste water. These systems will ensure that all of Greensburg's aquifer withdrawals are recharged with clean water and that site runoff and waste water discharges will be as pure as the rain that falls in Greensburg.

The last issue to be addressed in the first part of the master plan was that of economic development. Only after sufficient progress had been made to come up with sustainable clean energy, city-wide efficiency, and good urban design, would or should a strategy to recruit businesses be attempted.

The second half of the master plan was given to the City Council in May 2008. Steve Hardy made the presentation of Phase II to a newly elected City Council. Central to his plan was the notion of generating about four megawatts of electricity from wind. "Green power" off the grid would be purchased "when the wind is not blowing." Hardy also introduced

---

[60] According to Carolyn Szczepanski (*Pitch*, April 8, 2008), Greensburg had paid BNIM $286,000 for its services as of April 2008. That sum would swell over the next two years and become, itself, an issue in the town's reconstruction.

an idea popular in many European cities: Free bicycles to get about town in order to minimize vehicular use. This idea was adopted by Greensburg GreenTown.[61]

One BNIM speaker commented on the need for parks and green corridors. He touched on five recreational areas and talked about native and xeric plants—xeric plants being those which are not native to south-central Kansas but with water needs similar to those of native plants.

****

Ideologically Greensburg GreenTown was running in tandem with the City of Greensburg. Interaction was friendly and supportive, but there were few, if any, instances of concrete collaboration. The organization involved, at the outset, a board of seven volunteers—one of them was Jerry Diemart; an executive director—Daniel Wallach; and a secretary—none other than Alanna Goodman, who resigned her seat on the board to become its secretary. Greensburg GreenTown maintained a web page, and it announced plans to construct a dozen demonstration houses which would be built using environmentally sensitive techniques. The idea, first raised in the FEMA Long Term Recovery Plan, was to rent out these houses to eco-tourists. According to Wallach, "there is no science museum in the country that allows you to come stay overnight, but this one will."

Wallach was intent on developing Greensburg as a mecca of eco-tourism. "Tourists would get guide books and would be able to go around to the different sites and learn about the process of building green," he said. Buildings constructed using green methods would have what Wallach called "education stations." He was quoted in the March 17, 2008, edition of *Time*, as saying that Greensburg "could be a living laboratory to demonstrate to the rest of the country and the world what a town of the future could look like."

Rent money from the demonstration houses would be used to maintain existing structures and possibly to finance other green initiatives. The houses would be "upgradable" and new technologies would be displayed, kaleidoscope-like, one after the other.

The first of these houses was to be built on land donated by a Greensburg couple, Scott and Jill Ellers. Designed by Robert McLaughlin, this first dwelling was slated to house the Greensburg GreenTown offices, library and resource center.[62] Other houses were to be of straw bale construction. All the houses were so energy-efficient that their energy costs were estimated

---

[61] Although, like other projects, it came to nought. See the chapters *The Fruit of Green Labor* and *Aftermath of Compromise*.

[62] Things did not work out this way, but that comes later.

to come in around $700/year. But the parade of ecohomes was slow to take shape, delayed for a variety of reasons, the need to raise capital being the thorniest obstacle.

****

Throughout the fall of 2007, the move to go green gathered momentum. The Estes brothers—to cite one example—had to rebuild since the tornado had destroyed their farm equipment agency. Despite their fracas with Lonnie McCollum at the end of May, they opted to go green. They built outside the city limits of Greensburg, thereby evading the building code issues which had precipitated the dust-up with McCollum. But, like St. Paul on the road to Damascus, they had seen the light, a green light at that—as did the GM distributorship and the Baptist Church. The Estes brothers' decision to break green ground set a precedent for John Deere distributorships. "It will be the blueprint, the standard." Kelly Estes told Kevin McClintock in the *Signal*'s commemorative issue of May 4, 2008, that

> . . .their new facility, currently under construction, will have numerous innovative energy-saving devices, such as wind turbines and heat supplied by corn burners, run by oil from the tractors and combines. There will be more natural lighting. In place of asphalt and oil will be crushed concrete, a reusable natural resource. There will be a wind-powered aerator out on the pond and, to decrease light pollution, even a special light shining down on the 40-foot American flag out front.

John Deere's attitude evolved over time, starting shortly after the tornado. One crucial step of that development occurred in the late fall of 2007 at a steak restaurant in Pratt where the Estes brothers, Daniel Wallach, Chuck Banks (the State Director of USDA Rural Development), and Dave Jeffers, a manager from John Deere's headquarters in North Carolina, met for four hours to consider the implications of the distributorship's going green. This was the first time that Jeffers had set foot in the state of Kansas. Mike and Kelly Estes, with Dave Jeffers' support, stressed that sustainability needed to be compatible with getting the company up and running: Ecology was a secondary issue. But it was not impossible to persuade everyone that there was no contradiction between the need to rebuild quickly and environmental responsibility. When asked about the meeting 18 months later, Mike Estes said that the decision to rebuild green had already been taken. The Pratt get-together just "put meat on the bones." Not only would BTI John Deere come back green, but the dealership would serve as a prototype for John Deere distributors countrywide.

It is puzzling, given Wallach's prominence in the movement to take Greensburg green, that he and John Picard had little to do with one another. Wallach said of Picard that "he was not very accessible." Of course, John Picard did not move to Greensburg in 2007-2008. "He was always popping in and out," John Janssen said. In the same breath, however, the mayor described Picard as "a tremendous asset."

Among his other morale-boosting activities, Picard arranged for five Greensburg high school students to attend the November meeting of the US Green Building Council (USGBC) in Chicago.

The students raved about their Chicago experience. Planet Green was there to record their enthusiasm, filming an on-site (and staged) conversation between them and Picard. The students were later invited to make presentations at a community gathering on Wednesday, February 6, 2008. This was a meeting at which USGBC president, Rick Fedrizzi, also spoke. "You are the luckiest community on the planet with this kind of honorable and intelligent youth among you," he said of the students who decided to form a Green Club when they got back to Greensburg. One of the students, Levi Smith, said:

> The reason we formed the Green Club is that we are the leaders of tomorrow. I mean, that's just a fact.[63]

The line drew laughter from the audience, and John Janssen, noting the boy's puppy-like eagerness, told Levi that he would have to be dewormed.

Previously Levi, in concert with two high school comrades, had spoken at a town meeting held in the high school's temporary gym. In an *NPR* interview which aired on December 27, 2007, he said:

> Before the tornado, I was not going to come back. I was going to go to college and who knows where. This community was dying. Now, I'm definitely coming back. And I know probably a . . . majority of my friends are.

The Green Club which Levi helped start began projects which included maintaining recycling bins around the high school campus and starting a light bulb exchange, in which incandescent bulbs were turned in for fluorescent ones. Green Club students made the trip to Boston, one year later in the fall of 2008, to attend another USGBC conference. Levi Smith really liked Bean Town. And his association with Planet Green was reshaping his career ambitions: Levi began to consider a future in acting and life on the fast track in Hollywood.

---

[63] Mark Anderson, "Meeting offers reasons for optimism," *KCS*, Feb. 12, 2008.

****

All the cheer-leading was fine, said the skeptics, but what would going green cost? After all, an architect for MVP had already revised upwards Steve Hardy's estimates for the add-on expense of going green. Where was the money going to come from? And what, for that matter, was John Picard being paid for his civic cheerleading? In fact, his total compensation from the city of Greensburg was $25,000, a figure which came direct from Steve Hewitt.

Substantial sums had already been spent to clean up the debris in the wake of the tornado, to restore utilities and to redevelop business. Although organizations like the Salvation Army, United Way, and the American Red Cross spent millions of dollars on relief, the lion's share of the money spent on Kiowa County and adjacent areas came from three sources: the State of Kansas, FEMA, and other branches of the Federal Government.

Of the more than $35 million raised or dedicated to Greensburg projects by FEMA, more than one third was spent to bring relevant government agencies to town to help in the recovery effort. Well over another third was spent on debris removal, with the remainder spent on housing and other needs.

Seventeen million dollars was raised or dedicated by the State of Kansas. Of this sum, $5 million was approved for disaster relief. An additional $5 million was allocated to a Kiowa County Business Restoration Program with an offer of $2 million to businesses in sales tax exemptions. And in October 2007, another $5 million was approved for a Housing Assistance Program to provide forgivable mortgage loans.[64] The first company to avail itself of these funds was Manske & Associates of Wichita for its townhouse project.

Manske erected 32 townhouses on the site of the former school complex. The rents for these ran from $373 to $750 per month. They were income-dependent. Attention was paid to sustainability and energy efficiency: An operations manager for Nunns Construction, the builder for the townhouses, said that the walls had an insulation value of R-30. Wet cellulose and a one-inch thick foam board had been used to insulate the walls. Fluorescent bulbs and water-efficient toilets were installed throughout the townhouses. Their cost: $3 million. The townhouses which Manske built in Greensburg achieved LEED platinum certification, a first in the US for such a project.

---

[64]A forgivable loan is one made with the understanding that if the borrower meets certain conditions, repayment of the loan will not be required.

Of the first $60 million raised or dedicated by federal agencies for the town's recovery and rebuilding, $20 million was approved by the US Department of Labor to employ Greensburg residents for cleanup purposes. At the end of May, 2007, the Kansas congressional delegation, acting with unusual speed, succeeded in attaching a rider to a funding bill for the Iraq War which channelled $40 million to disaster-stricken Kansas counties, including Kiowa County. The money was under the control of the US Department of Agriculture (USDA) and was earmarked for expenditure when insurance and FEMA funding dried up. Of this money, $16 million was set aside for planning, housing, business redevelopment and community infrastructure and facilities. A further $10 million was designated as "unallocated/reserve" funds, to be used if and when additional needs arose.

Kansas State Director of USDA Rural Development was a man called Chuck Banks. He was the USDA representative at the meeting in Pratt when the decision to take BTI John Deere green had been hammered out. Given the role that Congress had bestowed on the USDA, Banks was one of the major players in the effort to secure funding for Greensburg development. He is a fifth-generation native of Kansas. The first of the Banks came to the state in the 1850s as part of an abolitionist colony which settled in a town called Wabaunsee. Under threat from pro-slavery marauders from Missouri, the colony obtained 27 rifles—and 25 Bibles—from their sponsoring church in Connecticut which was helped by a Congregationalist minister in Brooklyn, Henry Ward Beecher. (This preacher was the brother of Harriet Beecher Stowe, author of *Uncle Tom's Cabin*.) The fledgling congregation in Wabaunsee was known afterwards as the Beecher Bible & Rifle Church.

Chuck Banks had served for six years in the office of then-Congressman (later Senator) Pat Roberts before coming to the USDA in December 2001. Although his office was in Topeka, Banks spent much of his time from May 2007, to May 2008, in the vicinity of Greensburg. The week following the tornado, he went back and forth four times to Kiowa County.[65] Banks' first meeting with a FEMA official took place that week in Wichita and lasted a mind-numbing five hours. At times, Banks seemed omnipresent, talking at a variety of relief and recovery meetings, appearing as a guest speaker at a July 10, 2007, meeting of SCKTRO (whose construction project manager at the time was James Bond) and as a participant in numerous municipal gatherings in Greensburg. One of these was the so-called Big Tent Meeting which took place in Greensburg on May 11, 2007. There Banks confronted a sea of people, some on crutches, some in bandages. He described this as the most nerve-wracking public speaking event of his career.

---

[65] In the twelve months following the tornado, he averaged ten round trips a month between Topeka and Greensburg.

Funding in Greensburg involved public-private partnerships. The first concrete example of this was provided by the business incubator, a 9,200 square-foot facility designed by MVP with assistance from BNIM.[66] With space for service-type businesses on the second floor and a restaurant/deli on the lower level, the incubator would boast "rain gardens" and solar-reflecting concrete. Chuck Banks pledged $2 million from USDA for the building and announced that an additional $1 million from a company later identified as Sunchips, a subsidy of Frito-lay, had been secured.[67] Steve Hewitt, who had said of funding gaps that "all [they] do is create partnerships," was exultant. "Now that's what I call closing a gap!" he crowed.[68] The budget for the building, including site preparation, was estimated in a City Council meeting on January 22, 2008, at $2.76 million. So the USDA and Frito-lay donations should have sufficed to cover the expense of raising the incubator. But eventual cost overruns exceeded those donations. The difference was made up by contributions from Leo DiCaprio, who wrote a check for $400,000, and Craig Piligian, who ponied up an additional $200,000.

Banks was the right man at the right time for Greensburg because of his unswerving loyalty to the idea that renewable energy sources needed to be cultivated. "The road to energy independence for this country runs through rural America," he said. Kansas had been promoting green development for some time before the tornado. For Banks, it was both opportune and natural to dialogue with Greensburg.

To get a business incubator up and running as soon as possible was becoming increasingly urgent in light of the town's fumbling attempts to recruit industry. What progress that was made for several years after the tornado was made on paper. Example : Torsten Energy, a company headquartered in southwest Kansas, announced in early December 2007 that it would partner with Ball Industrial Services to build a biodiesel plant at the Greensburg industrial park.[69] "We still have a lot to do before we can actually break ground," said Patrick Stein, a representative of the company who would oversee the project. This, it turned out, was something of an understatement. "Terrific news for Kiowa County," exulted Dennis McKinney. "The more businesses that commit to building in Greensburg, the sooner hardworking families of our community will be back in town and back in the work force." But long after the announcement, no ground had been broken on the biodiesel plant (projected to employ 20 to 25 people upon completion). With the recession which hit the country in 2008,

---

[66]MVP had little, if any, experience in green building, a deficiency which BNIM was to remedy.
[67]Mark Anderson, "Meeting offers reasons for optimism," *KCS*, Feb. 12, 2008.
[68]Frito-lay became a sponsor of Planet Green's initial 13-part series on Greensburg.
[69]"Biodiesel plant picks Greensburg," *Lawrence Journal-World*, Dec. 4, 2007.

what looked at first blush like a delay began to look more and more like a cancellation.[70]

On the other hand, just two days before the anniversary of the tornado, it was announced that the Economic Development Administration (EDA) of the US Department of Commerce would provide almost all the funding to repave Main Street and rebuild sidewalks along four blocks of that thoroughfare, for $2.3 million.[71] A check was presented to John Janssen the day before the first anniversary of the tornado. Only four weeks had elapsed since Steve Hewitt had first discussed the streetscape project with the EDA. With pedestrian traffic in mind, the new design called for narrowing the original Main Street to increase the width of the sidewalks by six feet. Native plants and mature trees figure in the BNIM master plan for the street as well as the use of rainwater to sustain them.[72]

On Tuesday, January 22, 2008, Nye and Associates, a marketing consultant with experience in working with a town devastated by a tornado[73] pitched a plan to the City Council to promote Greensburg. It included the production of a brochure and a DVD targeted at business audiences as part of a public relations program, a website update, and a direct mail campaign. All of this was to be incorporated in telling what Nye called the story of the "Great Greensburg Comeback." The Council liked what it heard, but it wanted to work with someone on-site, not an Emporia-based company. It eventually hired just such a person.

While town authorities were working to implement a strategy of LEED certified rebuilding, private citizens were rebuilding on their own. As of December 2007, Mayor John Janssen had granted 664 requests for building permits.[74] "What we're hearing from people is that they don't want architecture that reflects the turn of the last century, but buildings that reflect this century," BNIM's Stephen Hardy was quoted as saying. "Greensburg, like much of the rest of rural. . .America, was headed on a long path of decline before the tornado, so residents want to rebuild the right way."[75] Be that as it may, much of non-public rebuilding was decidedly conventional in appearance although residents were more sensitive to the need to conserve energy by insulating more effectively and installing efficient heating and AC systems. The guts may have been green, but the skin was modelled after

---

[70] See the chapter entitled *Aftermath of Compromise*.

[71] Mark Anderson, "U.S. Commerce Department promises $2.3 million to rebuild Greensburg's downtown," *KCS*, May 9, 2008.

[72] In a sign that townsfolk were already out of sync with the utopian ideals pouring forth from BNIM, Rex Butler found fault with the narrowness of the new Main Street. "Not enough room to turn a pickup around," he groused. Rex's opinions carried weight, particularly after his election to the City Council in 2010.

[73] This was Pierce City, Missouri.

[74] Michael Burnham, "Green Building: Tornado-flattened town plots rebirth as wind-power capital," *www.bdcnetwork.com*.

[75] *Ibid*.

what peoples' parents had lived and worked in. The *Kansas City Star*'s Aaron Barnhart made the observation that:

> I found it interesting, during my visit, that the Mennonite Disaster Service in Greensburg wasn't sending volunteers to build new structures but to help fix up the few that are still standing. Simple lifestyles don't mesh well with a broader culture that's urged to consume, almost out of national duty.[76]

It should be kept in mind, however, that one (Holdeman) Mennonite builder, Lloyd Goossen, used Insulated Concrete Form (ICF) blocks in his structures. In this technique, which was developed relatively recently, the components are put together like Lego blocks. The resulting walls have an insulation value of R50. Goossen gets his ICFs from the Canadian province of Alberta, to the west of his native province of Manitoba. Despite his use of this contemporary building technique, the elevations and floor plan of Lloyd's new house—apart from the generous dimensions of some of his redwood-framed windows—resembled the ranch-style houses which proliferated on this continent in the second half of the twentieth century.

There were exceptions to the rule, however. A Lawrence-based design/build firm, SIPsmart Building Systems—SIP for 'structural insulated panels'—planned a decidedly non-traditional house for Jill and Scott Eller. This was the couple who had contributed land to Greensburg GreenTown for the first of its eco-homes.

"The Ellers' house got swept away, and now they're coming back with something completely different," said Jon Red Corn, a young architect who works for SIPsmart. Jon, who owes his surname to Native American ancestry—his father's side of the family has its roots in the Osage tribe of Oklahoma—is a colleague of SIPsmart owner Michael Morley, who started the company under the name "Morley Builders" in 1990. Morley was the architect for the Ellers' new house. It featured a geodesic dome (à la Buckminster Fuller) split in half and flanking a central core. Typically built with equilateral triangle panels, the geodesic dome halves of the Ellers' house used right-triangle panels which were easier to handle. The half-domes went up one month before the central core of the house. It wasn't long before the house became known in town by nicknames which ran the gamut from "Boob House," to the "Princess Laia House."[77]

****

---

[76] Aaron Barnhart, "The Greening of Greensburg," *Kansas City Star*, June 15, 2008.

[77] A previous point of contact: The Ellers had once lived in the abandoned Methodist church bought by Janell Sirois. That was the building which Charlie Jones stumbled across on his odyssey around Greensburg the night of the tornado.

Lynn Billman, a Senior Analyst with the National Renewable Energy Laboratory (NREL), first arrived in south-central Kansas in July 2007. She came to give presentations on energy conservation in private homes. Colorado-based NREL, which is part of the Department of Energy (DOE), set up shop in a FEMA trailer. There a team of ten people began offering technical advice on energy efficiency in all types of buildings. The summer months, according to Billman, were chaotic. "There were too many chiefs and not enough Indians," she said. Billman obtained a B.S. in chemistry from UC-Berkeley before going to work for Chevron. Subsequently she did financial corporate planning in San Francisco before moving to Colorado to work for what was then called the Solar Research Institute (SRI) as a science writer. SRI became NREL in 1991. She worked for the lab from her first days in Colorado.

Asked if Kansans cared to hear what NREL had to say, Lynn Billman replied that her reception had been very positive. She met Daniel Wallach in August 2007, and the two had hit it off. Wallach opened doors for NREL which began to collaborate with Greensburg GreenTown. When FEMA left in August 2007, access to her audience of town residents became easier. She met with builders to show what could be done in an integrated sense, how buildings could be designed for energy efficiency, for example. In the case of high profile buildings, like the Greensburg school complex or the Kiowa County Hospital, computer modelling came into play.

"There's been remarkable progress here," observed Billman. She used to commute to Greensburg from Colorado several times a month. She stayed in touch with her family—a husband and three grown children—by cell phone.

The Colorado-based lab got high marks from Wallach:

> NREL did it right. They operated out of a presence rooted in the community. They supported the community in planning how they wanted to implement sustainability, a very empowering approach.[78]

Not all organizations got such praise from Wallach. The worst, in his opinion, came to town with the attitude "We're here to fix you and make you better if you do as we say."

> It was a real patronizing attitude. . . I remember one of the people from a government agency told a town meeting after being here several months, 'We're going to take the training wheels off now and leave you on your own.'

---

[78]Mark Anderson, "Little green man around town not from Mars," *KCS*, Apr. 17, 2009.

Wallach sensed a parallel between the disruption of Greensburg life and the ten-year battle which he and his wife had waged against their illnesses. He also saw a parallel between his own experience with doctors and what he heard from some of the agencies which came to Greensburg after the tornado. The best, in both cases, took the attitude that "you have everything it takes to be whole again, and we want to partner with you in that endeavor."

****

An example of the progress claimed by Lynn Billman was given by Jerry Diemart, who renovated a house on North Sycamore St. The building had an ecological element which had grown out of some work which Jerry had done while he was employed at the Texas Women's University: A load-shedding mechanism which used a programmable logic circuit to shut off some electrical appliances, like a washing machine—something which could be restarted automatically without detriment to its operation—when the building's air conditioning was turned on. When the AC kicked off, the washing machine would be reactivated. With this sort of "power sharing," the wattage of the house's electrical supply could be cut back. Diemart reckoned that the electrical capacity of the house's wiring could be kept down to eight kilowatts.

The renaissance of the town, in the hands of private citizens like Diemart, however, came with its share of birthing pains. Unscrupulous contractors took advantage of their clients. In the typically blunt way he expresses himself, John Janssen said that one firm from Wichita had "screwed everybody." But people were slow to make formal complaints. Said the town's Community Development Director, "I think they're afraid to admit they made a mistake. . .[and that a contractor] would turn on them when they're three fourths finished with their house," if they filed a formal complaint. He told the *Signal* that his office had received "phone calls and informal comments on the performance of two or three contractors," but he had had only one person "actually sit down with me one-on-one and give me a detailed complaint."

The best known dispute between a Greensburg couple and their contractor pitted Marilyn and Robin Brown against an Oklahoma-based home remodelling business called AAA Renovations. It was owned by Al Soulek. The Browns alleged that Soulek had left them high and dry with a partially finished house—their "dream house" in Marilyn Brown's words—after absconding with $181,000 of the Browns' money.[79] The Browns had retained Soulek after Marilyn's mother, who attended the same church as Soulek in

---

[79] Mark Anderson, "Couple's rebuilding effort interrupted by absent contractor," *KCS*, Dec. 4, 2007.

a suburb of Oklahoma City, recommended him to them. One of Soulek's checks to a local purveyor bounced. The Browns became uneasy. The Kiowa County Sheriff's Office ran a check on Soulek and uncovered some "red flags" on AAA Renovations. But by the time this was discovered, the Browns had signed a contract with Soulek and handed over a sizeable chunk of cash to him.

Soulek had been prosecuted in 2006 by the Oklahoma Attorney General's office on six counts of home repair fraud, a charge that was dropped on the condition that Soulek avoid charges of fraud in the future (something which he failed to do).[80] The Browns' accusations "landed the builder back on the court docket" in Oklahoma. Their lawyer admitted that Greensburg was "between a rock and a hard place," because the town was torn between an impulse "to encourage builders to come in" and an unwillingness to bring in those "who like to take the money and run."[81]

The Browns' problems became known nation-wide because they were played up on an episode of the Planet Green series. An investigative reporter who ran a program called "In Your Corner" on Channel 4, Oklahoma City, alerted the Browns to the imminence of Al Soulek's trial in Oklahoma on charges of fraud unrelated to the Browns' case. The Browns decided to attend the sentencing. Planet Green was quick to follow, actually filming part of the proceedings through a glass panel in the courtroom door.[82] Marilyn wanted to look Soulek in the eye. At one point, Robin actually stood up in the courtroom and stared at the contractor, arms crossed on his chest. Soulek had previously pleaded guilty to six counts of fraud. At the sentencing on May 14, 2008, he was told that he would spend 90 days behind bars and that he was being ordered to pay in excess of $44,000 in restitution. It appears, however, that because of overcrowding in Oklahoma jails, the convicted contractor spent only two days in the pokey.

Two Methodist churches in the Kansas City area—Grace Methodist and Woods Chapel—came to the Browns' succor, helping them to complete their house in Greensburg. The Browns were advised that Soulek was broke and that they would be lucky to get $5,000 out of the contractor. But a negotiator in Pratt wrung a check for $10,000 out of AAA Renovations in Novembr 2008—not, however, before two of Soulek's personal restitution checks had bounced.

There were complaints about another contractor. Two of the parishioners at the Lighthouse Worship Center—the church which Vernon Davis attended—were "blessed to have work done on their houses by James

---

[80] Mark Anderson, "Brown presses for action against contractor," *KCS*, Jan. 3, 2008.
[81] *Ibid.*
[82] This earned Planet Green the threat of a suit by Soulek's attorney, a threat which never resulted in any action.

Bond," according to Pastor Christa Zapfe. But there were other, less fulsome reports on the contractor's work. One of the projects at which he was interviewed for this book was a house for Pat and Charlie Jones. Charlie Jones claimed that the work on the walls of the house did not comform to the manufacturer's specs insofar as only one rebar was used where there should have been two overlapping rebars. Charlie produced a building inspection report done by D & B Engineering, a Wichita firm (dated June 12, 2008), which stated that "photos clearly show the crooked walls and unlevel floor." Furthermore "the . . . unlevelness and waviness implies that little if any bracing was provided."

The SCKTRO Board of Directors offered to use steel plating and bands to correct the problems in their house. The Joneses eventually agreed to such an arrangement. They were able to occupy their house on May Day 2009, even before the exterior work had been finished. A month after the Joneses had reoccupied their house, Charlie walked in his front door one day, grinning broadly and holding in one hand a scruffy-looking shoe.

"This was David Lyon's shoe," he said. David Lyon was one of the eleven victims of the tornado and a neighbor of the Joneses. Charlie had stumbled across his flashlight on the night of the tornado, unaware that Lyon's body lay only a few feet away. Charlie was exultant about his discovery:

> I just found it on my walk. I can tell it was David's because it's missing a shoe lace. David never used laces. I may go back and look for the other shoe.

In late April 2009, another couple voiced a complaint about the work done for them by James Bond. Because they wish to remain anonymous, they will be called here Mr. and Mrs. Smith.

What was their beef? James Bond laid the concrete footings for their house. Although Bond was working for SCKTRO, it was his own company in Junction City that billed the Smiths $15,172.37 for this work. They paid it by a bank draft (dated September 7, 2007). Time passed without SCKTRO volunteers reappearing on the work site to continue construction. Anxious to make progress on their home, the Smiths engaged a builder from Arizona to finish the job. While the Arizona crew was working at the Smith site, a couple of Bond's volunteers dropped by to ask what the deal was. The Smiths sent them packing. On December 19, 2007, the Smiths paid a local firm, Heft & Sons, $11,116.93 for the concrete used by the Arizona company to do their flatwork. In the spring of the following year, they were surprised to receive from Heft & Sons another bill for $6098.28 for the concrete used in James Bond's footings. In addition to their concrete bills, the Smiths received bills for two separate shipments of rebars, the first to pay for 150 rebars ($1,027.50), the second for twice that number ($2,235.02). A friend who saw the invoices asked the Smiths exactly how many houses they were building. They insisted that none of the rebars involved in the second bill were used in their house. Confusion may have arisen because James Bond was using the work site at the Smiths' as a kind of dumping ground for supplies for various properties his volunteers were working on.

Were these screw-ups a case of incompetence? Sloppy bookkeeping? What? The plot thickened when SCKTRO presented the Smiths early in 2009 with a bill for approximately $20,000. The Ministerial Alliance, rather than the Smiths, paid this last bill.

James Bond left South Central a matter of days after being interviewed for this book. That was at the end of March, 2008, several months before what was expected to be his final days in Greensburg. The Smiths catered his going-away meal, in fact, serving barbequed ribs, baked beans, and cold slaw.[83]

What's the final verdict on James Bond and what he did in Greensburg? The town is divided in its opinion. One comment from several people who had dealings with him was that James Bond was "over his head" with the work he was doing in Greensburg.

"James could never say no," one town minister said. "If someone came to him with some kind of issue, he would say 'No problem. I'll handle it.'"

The same minister expressed concern over Bond's bursts of anger. The contractor lost his temper on several occasions. A couple of these outbursts were spectacular. One of these led to his being furloughed for a couple of weeks to calm down.

Not all the criticism went the contractor's way, however. One resident commented sharply that "it aggravated me no end that people accepted volunteer work, then complained about it." There was a big difference, after all, between benevolent and contract work. All James Bond had was volunteer labor.

The contractor had little to do with Greensburg after leaving at the end of March, 2008. He did phone a former colleague, Matt Deighton, a few months afterward to say that he wanted people to know that he was thinking of them. He had just watched a TV program about the town,

---

[83] Another foodie footnote: In the early days of the relief effort at Greensburg, James Bond had treated New York fire fighters—who had come west to help with the cleanup—to $80 worth of boneless pit ham, again provided by the Smiths. Bond told them, in a rather cavalier way, to put the ham on his tab. According to the Smiths, this was a chit he never redeemed.

presumably as part of the Planet Green series. Bond said that he was phoning from Washington, D.C. The months clicked by. Bond dropped into town at the end of 2008 to retrieve a trailer and some tools. It was then that he came by the SCKTRO office. Matt Deighton was not wearing his glasses—he took them off when he worked on the computer—so the image of James Bond, fuzzy at first, came into focus slowly as the man approached Deighton's desk.

"Hi, Matt," said Bond.

"Hi, James," said Deighton.

The two men shook hands, and James Bond left.

Deighton was not one of James Bond's fans. He (Bond) spent time, according to Deighton, in "coaching people to become part of his cult." On the other hand, one resident whose house had been put to rights by Bond's volunteers said of the man that "he had a heart of gold." As he drove off into the sunset, James Bond left, for some, as the Lone Ranger, and, for others, as Elmer Gantry.

Reached by cell phone years later, James Bond said that he had taken up work with World Hope International after leaving Greensburg. As a disaster response manager, he had worked in Iowa at the time of the floods in that state; in Texas after hurricanes struck; and in his home town of Manhattan, Kansas, when a destructive tornado passed through there. Budget cutbacks cost Bond his job.

What did he have to say about the criticism of his work in Greensburg? Bond was not aware of any problems although he acknowledged that he knew of a house owner who had complained about the insufficiency of rebars in his walls. But SCKTRO had found the complaint to be without foundation. Bond was ready to stand behind all the houses that he had built. Some were even overbuilt.

****

For the three years following the tornado, Greensburg never completely dropped off the media's radar screen. Its decision to achieve LEED certification for the reconstruction of its public buildings is one of the reasons it continued to bask in the environmental spotlight. A LEED resolution, drafted in cooperation with BNIM's Steve Hardy, was adopted by the City Council at its December 17, 2007, meeting. It was a breathtakingly bold measure which called for construction of City Hall, the business incubator, and a tourism center at the Big Well to meet LEED's platinum certification. All buildings with a footprint exceeding 4,000 square feet would need 1) to reduce energy use 42 percent over building code requirements, and

2) to utilize renewable energy sources like wind. Undeniably Greensburg had an abundant supply of that commodity. "The city is sitting on an excellent wind site," Steve Hardy said. "We're trying to figure out a way to capture that resource, maybe [with] an industrial wind farm near town or small turbines throughout town."[84] Lynn Billman reckoned that the wind in this part of Kansas was "a constant wind. . .[possibly] 16 mph at 50 meters above ground. That [kind of] moderate-strength wind is your best economic source for wind power."[85]

Steve Hewitt told those in attendance at the December 17 meeting:

> We need to realize this is unprecedented because there's no city that's passed a platinum resolution. Some gold, yes, but not platinum. This is ground-breaking stuff.[86]

Interviewed on *NPR* in December 2007, Hewitt maintained that:

> There are only 14 platinum buildings in the US. When all is said and done, I'd like to have four or five here in Greensburg. And don't tell me that doesn't put us on the map when people are taking a hard, serious look at this community. I think it does.

Hewitt and Hardy expressed the hope that individuals rebuilding their homes and businesses would follow the city's lead. There were indications that their hope, at least in part, would be realized. As mentioned above, the Manske townhouses were built to LEED platinum certification. Superintendent of Schools Darin Headrick—one of the members of Lonnie McCollum's Rat Pack—was confident that the rebuilt high school would meet the city's platinum standard. "We can make sure our school and its design is one of the unique reasons as to why people would want to live here," he said. Others were less ambitious: Kiowa County Commissioner Gene West said that MVP was working on designs to rehabilitate the courthouse that had been damaged but not destroyed by the tornado. In his opinion, "silver is doable, maybe gold. . .but platinum isn't."

Cost estimates of achieving LEED certification varied. Steve Hardy put the upfront construction cost increases at around 6.8 percent. But Mark McCluggage of MVP gave much higher percentages. According to him, LEED certification could tack on as much as 30 percent (in the case of platinum) to upfront costs.

---

[84] Michael Burnham, "Green Building: Tornado-flattened town plots rebirth," *www.bdcnetwork.com*.
[85] "All Things Considered," *NPR*, Dec. 27, 2007.
[86] Mark Anderson, "Council takes the LEED with new building," *KCS*, Dec. 19, 2007.

The February 6, 2008, meeting was a celebration of sorts to commemorate the City Council's decision to build to LEED platinum certification. Three hundred people crowded into the high school gymnasium to hear a number of distinguished speakers. First up was Bob Berkebile of BNIM who paid tribute to the town's determination to rebuild in a sustainable way. "This healthy virus is beginning to infect private concerns outside this community," he observed.[87] Not only the Shank Motors (GM) car dealership but the Estes brothers' BTI John Deere dealership were rebuilding to meet LEED's platinum certification—the first, for that matter, in the nation.[88] In fact, the decision to require platinum certification for its public buildings put Greensburg at the forefront of the "greening" of the nation.

"This is about setting a new model for how we build communities and economies in the 21st century," Berkebile told his audience. "You're on the verge of becoming the first carbon-neutral community in America."

Berkebile went on to announce plans for what he called an "Essential Technology Center" in an "eco-industrial park," to go up near Greensburg where a process to convert solid waste to produce energy without combustion would be used. He then introduced Steve Hardy and another colleague from BNIM, Rachel Wedel, who presented a short video that showed events from the tornado to the recovery effort. John Picard figured in the film as well as Daniel Wallach and Governor Kathleen Sebelius.

John Picard then took the floor to lather the crowd with more compliments. He described the recovery effort as one "to go into the world to protect nature and life-giving forces."

He went on to look directly at Steve Hewitt and say "Steve, you are a hero." Then, directly to the crowd in attendance, he said, "All of you are heroes." Both lines were received by what the *Signal* described as "deafening applause." Were that not enough, Picard added that he often felt less than complete in his life but that trips to Greensburg left him feeling renewed and invigorated. "When I go home from here, I am completely happy."

In his part of the program, USGBC's Rick Fedrizzi told people how stunned he was at the courage they had shown. "Rather than just putting a roof over your head as quickly as you can, you've taken a deep breath and looked ahead to the future." Environmentally aware cities like New York, San Francisco, and Seattle had not made the same level of commitment to LEED standards as Greensburg had.

---

[87]Mark Anderson, "Meeting offers reasons for optimism," *KCS*, Feb. 12, 2008. The quotes which follow all come from this source.

[88]The GM dealership was rebuilt with a number of green features, but a decision was made not to seek LEED certification.

To conclude his part of the show, Picard introduced eleven colleagues who had flown in with him from California. Picard gave only their first names and the businesses they were involved with. For example, "Eric" was with "one of the world's largest commercial development companies." Another was with "one of the biggest energy companies in the world" while another was with a "huge water treatment company." All could potentially contribute to the rebuilding of Greensburg. To wind up, John Picard unleashed his gift for the theatrical by claiming, "We're going to rebuild this town green and then we [shall] rebuild this whole planet green."

As bizarre as it might seem to disguise the identities of his fellow-travelers, Picard was aware of the need for anonymity. A premature announcement on NPR had cost Greensburg its chance to acquire a wind-powered Google data center. It had been planned to bring Picard's companions to Kansas by a decidedly non-green Lear jet. But they had actually flown into Wichita on a commercial plane and driven the 100 miles to Greensburg by bus. John Janssen and Steve Hewitt had met and talked to them, no attempt to hide their names or company affiliations had been made, and Janssen and Hewitt could vouch for the authenticity of these distinguished guests. Still. When this book was finished, four years after what proved to be John Picard's last hurrah, none of these nameless movers and shakers from the West Coast had made a quantifiable commitment to Greensburg.

MEDIA MATTERS

"I want you to fly with me to Kansas," said Dorothy.

But the Monkey King shook his head.

"That cannot be done," he said. "We belong to this country alone, and cannot leave it. There has never been a Winged Monkey in Kansas yet."

–Wonderful Wizard of Oz

The National Guard coordinated press coverage of post-tornado Greensburg. Adjutant General Tod Bunting and his Director for Public Affairs, Sharon Watson, arrived there early on the morning of May 5, 2007. Bunting was there to direct the initial clean-up of the town, Watson to bring some kind of order to the press scrimmage which inevitably ensued. Watson had contacted the national media before going to the site of the disaster. When she got to Greensburg, however, she found only local media like the *Wichita Eagle* on site. Two public affairs officers from the 184th Air Refuelling Wing at McConnell Air Force Base outside Wichita were called in to assist. National media—including CNN, ABC, NBC, and the AP—began to arrive in Greensburg in the early afternoon of Saturday.

Early on, a decision was made to move media into the center of Greensburg. Reporters and photographers were demanding access to the town, anyway, but they were restricted to a few blocks between the grain elevators—which were still standing—and a line one block north of US 54. Media were escorted out at night and returned the following morning, sometimes as early as 4 a.m. Initially their interest focused on the body count in town, search and rescue missions, including the use of a canine corps, and the work of the first responders.

When Governor Sebelius gave her Sunday afternoon press conference, criticizing the federal government for sending to Iraq personnel and equipment that might otherwise have been used in Greensburg, media attention shifted to the Middle East and the deployment of the National Guard there.

Watson found herself fielding calls from newspapers like the *Los Angeles Times* and the *Chicago Tribune*. All manner of questions came in: Was the air safe to breathe? Was there a lot of asbestos lying around? Her previous experience in radio broadcasting and in reporting for the *Olathe Daily News*, where she had worked for ten years, gave her insight into what the media would be asking. Before coming to work for the National Guard only seven months before the Greensburg tornado, Watson had put in a 5-year stint as Communications Director with the Kansas Department of Health and Environment. Rumor control became a sizable part of her job.

****

The attention that the national media paid to the tornado had a brief half-life, but during the few days it lasted, some poignant vignettes were penned. Examples from *The New York Times* and the *Wichita Eagle* were cited in an earlier chapter. Another example from the *Wichita Eagle* focused on the leading tourist attraction of Greensburg, the Big Well. On Sunday, May 6, 2007, the newspaper ran a wistful reminiscence by Phyllis Jacobs Griekspoor in which she wrote:

> I've always thought of the well as kind of a monument to the persistence, maybe even the downright stubborn streak, that runs down the backbone of Kansas.

Two days later, the paper wrote that divers from the Sedgwick County (Wichita) Fire Department had discovered that the well had survived. Two divers descended 100 feet into water which was 10 to 12 feet deep. "All we found are a ball cap and a whole bunch of coins," one communicated through a walkie-talkie. The article concluded with the remark:

> The work crew then secured the staircase for safety reasons with lumber, so no one would accidentally fall into the well. To barricade the well, they used an old door with a poster that read "Kansas: the real experience." [89]

---

[89] On Tuesday, May 8, 2007, the Associated Press reported that a volunteer had dug through rubble to unearth the 1,000-pound meteorite which had been stored at the Big Well. It was moved to Hays before being returned to Greensburg at the beginning of 2010 and housed in the newly opened City Hall.

On Monday, May 7, 2007, the *Eagle*'s blog ran a story about the legendary Greensburg soda jerk, Dicky Huckriede. A co-worker at Hunter's Rexall Drugstore was interviewed in her Wichita hospital room. "I don't know if he made it," she said.

The *Eagle* reported shortly afterwards that Dicky Huckriede was safe and sound in Mullinville. The cardboard cutout of the famous soda jerk, which Erica Goodman had taken with her to Garden City, ended up in safe keeping with Ed Schoenberger, the town historian and cemetery sexton.

****

President Bush's visit to Greensburg on Wednesday, May 9, 2007, marked a turning point in media interest in Greensburg. When Bush left, so did much of the national press corps. The *Wichita Eagle* described it this way:

> Greensburg is a very different place today [Thursday, May 10, 2007]. Now that the president has been and gone, the national media have packed up and left, too. . . . What used to be downtown Greensburg is eerily empty.

The president's concern for Greensburg helped to keep the nation's attention focused on the town. His May 9 visit was the first example of presidential solidarity. There would be others. Why did George Bush lavish so much attention on a town with a population of 1400? Possibly this came as a counterbalance to the appallingly inadequate response to Hurricane Katrina in 2005. Greensburg is also part of the Republican base in this country. The president's advisors may have urged him to shore up that base by making a photo-op show of solicitude for Greensburg. Whatever the reason, Bush's visits to the town were appreciated. They offered a ton of publicity for Greensburg's struggle to reinvent itself. That struggle—the town's determination to rebuild green—kept the attention not only of the national media but of the international press as well.

Another example of the president's involvement in the community came in the form of an invitation to Steve Hewitt about two weeks after Christmas, 2007. The White House staff, which had kept its eye on Greensburg ever since the tornado, called to tell the City Administrator that President and Mrs. Bush were inviting him to attend the president's State of the Union address the following month. The Greensburg City Council picked up the tab for his air fare and hotel bill. Upon his arrival in Washington, D.C., a city which he had never visited before, Hewitt made his way to the White House where he was escorted into a reception where the Bushes' 15-odd guests were assembled. This included the former senator from Kansas,

Bob Dole. From the White House, the guests motorcaded to the Capitol where they were taken to the presidential box. Bush did not mention Greensburg that day, but Governor Kathleen Sebelius, in her Democratic response, did. "Greensburg will recover [due to] Steve's efforts and hundreds of others in our state and across the country," said the governor. She went on to say:

> But more than just recover, the Kansans who live in Greensburg are building green, rebuilding a better community for their children and grandchildren, making shared sacrifices and investments for the next generation.[90]

Sebelius urged Bush to move in the same direction:

> The majority in Congress are ready to tackle the challenge of reducing global warming and creating a new energy future for America. So we ask you, Mr. President, will you join us?[91]

A more substantive invitation came Hewitt's way in June 2008, when he was asked to testify before a Select Committee of the House of Representatives. Chaired by Congressman Edward Markey (D-Massachusetts), the Committee on Energy Independence and Global Warming wanted Hewitt to explain Greensburg's goals to them. Why was Greensburg aiming for LEED platinum certification? It was a goal which would cost the town as much as 20 percent more than conventional construction. Was this a wise way to spend taxpayers' money? Hewitt told the Select Committee:

> Greensburg now has the opportunity it never had before. I believe, as many around the world do, that we have reached a tipping point on the environment. . . . I am convinced that if we take action, and become environmentally smarter, we can shape the environmental and economic futures of all of us. Green starts in rural America.[92]

By deciding to go green, Greensburg had a fighting chance to survive. And the town would now be getting a congressional stamp of approval: Hewitt's eloquence led to the committee's support for additional USDA funding. A letter to Secretary of Agriculture Ed Schafer (dated July 15, 2008) stated in part:

---

[90] Mark Anderson, "Sebelius, not Bush, acknowledges Hewitt," *KCS*, Jan. 30, 2008.
[91] *Ibid.*
[92] Transcript of Steve Hewitt's testimony before the Select Committee on Energy Independence and Global Warming, U.S. House of Representatives, June 18, 2008.

The Select Committee requests that USDA recognize Greensburg's role as a national model for energy-efficient rebuilding standards, and to support its associated costs and environmental benefits. . . .

Building to higher efficiency standards is crucial to reducing global warming emissions, reducing the risk of the associated severe weather, floods, hurricanes and droughts. . . Replacing an inefficient building with another inefficient building makes no long-term sense, financial or otherwise. The Select Committee urges USDA to support Greensburg's forward-thinking rebuilding and energy efficiency efforts and looks forward [to] your response in a timely manner.

****

The interplay between what was actually happening in Greensburg and the way the media—TV in particular—portrayed what was happening played a significant role in the rebuilding of Greensburg. This is illustrated by how Planet Green's series on the town projected the town to the American public.

The result of Eileen O'Neill's huddle with Steve Hewitt was an agreement that allowed Discovery Communications—which reportedly invested $50 million in its spinoff network—to launch the fledgling Planet Green with a 13-part series on the town. "When we first partnered with the town, we were there because they declared they wanted to go green," O'Neill told the *Kansas City Star*.[93] "They saw the value of having ongoing media attention through our involvement."

Once this agreement had been secured, Leo DiCaprio agreed to collaborate on the series. It was reported that DiCaprio's production company was doing its own environmentally-centered reality show entitled *E-topia* which would shadow a team of construction workers, urban planners and environmentalists as they did a green make-over of a town. The tornado of May 4, 2007, produced a subject tailor-made for that kind of show.

DiCaprio opens the Greensburg series with a brief statement about the importance of sustainable development. Planet Green engaged Craig Piligian's *Pilgrim Films & Television* to produce the Greensburg series. One reporter had this to say about the involvement of DiCaprio and his production company:

---

[93] Aaron Barnhart, "The greening of Greensburg: Coming soon to a Discovery-owned channel near you," *blogs.kansascity.com*, Feb. 26, 2008.

Mr. Piligian pitched Leonardo DiCaprio, who came on board as production partner (with his production company Appian Way) and executive producer (along with Mr. Piligian). 'We knew that Craig and Leo had partnered to bring a project to TV, something that advanced Leo's amazing work in the area of environmentalism,' says Planet Green executive VP Eileen O'Neill.[94]

In a telephone interview conducted by Gavon Laessig for *Lawrence.com*, Planet Green executive producer Timothy Kuryak, also commented on the roles of Leo DiCaprio and Craig Piligian:

'Greensburg' came about through Craig Piligian. . .and Leonardo DiCaprio [who] were looking for a project to do together. We . . . approached them and said, 'Hey, instead of creating a town from scratch, why don't we document the rebuilding of this town that's going green?'[95]

DiCaprio sent out an e-mail message, quoted in the *New York Times* on June 10, 2008, which stated:

We are in the Environmental Age whether we like it or not. Unfortunately, our government has failed to respond to this monumental issue in the way they should. Until we become less reliant on foreign oil and put aggressive environmental policies into action, it will be towns like Greensburg, Kansas, taking matters into their own hands on a grass-roots level.[96]

Craig Piligian had made a name for himself in TV news at CNN and ABC. He left to make some of the first reality shows on TV. In 1996, he founded *Pilgrim Films & Television* and began to produce documentaries for the Discovery Channel including "Covert Action" and "CIA Files." Late in 1999, he began producing the "Survivor" series for CBS. Piligian has a personal connection to Kansas. His wife is a native of Hutchinson where the couple own a house. They were actually there on the night of the tornado. Interviewed by Aaron Barnhart in May 2009, Piligian had this comment to make about the Greensburg series:

---

[94] Debra Kaufman, "From the Ground Up," *Television Week*, Sep. 3, 2007.
[95] This quote is contained in Laessig's article, "Greensburg Returns," which appeared in *Lawrence.com*, in its edition of May 27-June 2, 2008.
[96] Kathryn Shattuck, "Playing a Leading Role in the Ruins of a Tornado," *NYT*, June 10, 2008.

You don't have people sniping at each other, backbiting, all those things realty shows play off of. We've done the reverse.

Piligian's executive producer on-site in Greensburg, Kathryn Takis, had worked on "Laguna Beach—The Real Orange County" for four years, "from beginning to end," as she wrote in an e-mail. "'Greensburg' and ' Laguna' are similar in the sense that both series followed the lives of people, but that's really the only similarity."

The Greensburg series was "at the center of what we want to do with Planet Green," said the president and CEO of Discovery Communications, David Zaslav, who added:

> We're the number one non-fiction media company in the world, but we also want to make a difference.[97]

For her part, Eileen O'Neill said of Planet Green that "the network is not only not finger-wagging, it's sexy, it's interesting, it's irreverent."[98] What Planet Green was offering its audience was "ecotainment."

There was, in fact, some apprehension about what the Planet Green editors and story producers back in Sherman Oaks, California, would do with the material their colleagues on the ground in Kansas were sending them. People involved in the rebuilding of Greensburg fretted that the series would need to create drama and conflict to retain viewers' interest. This would come at the expense of accuracy and an understanding of what the town was trying to do. BNIM architect Ruth Wedel put it this way:

> It's worrisome because they've gone into people's homes, and in a town of 1,500 people, everybody knows everybody. . . . The [townsfolk] would have gotten a certain amount of [attention], anyway, because of their progressive goals and tenacity in reaching them. . . . When you have a force like the Discovery Channel promoting your good deeds, you can't help but be glad.[99]

The Discovery filming crews got generally high marks from the citizenry. "The residents of Greensburg have welcomed us with open arms," wrote Kathryn Takis in her e-mail. Timothy Kuryak gave Gavon Laessig a more nuanced answer:

---

[97] Natalie Finn, "Leo Lands on Planet Green," *eonline.com*, 2008.

[98] David Bauder, "Planet Green not your father's TV channel," *www.msnbc.msn.com*, June 2, 2008.

[99] Gavin Laessig, "Greensburg Returns," *Lawrence.com*, May 27-June 2, 2008.

Like any person would be at first, they were reticent. They had been through a terrible tragedy and were trying to get their lives together. So on top of that, to have camera crews in your face isn't the most welcome thing.

Both Takis and Kuryak were in agreement on how Greensburg residents and the TV film crews eventually came to terms with one another. Said Kuryak:

They [the Planet Green crews] have really grown to love it. I've been out there several times and they've been embraced by the townspeople. They're almost considered townspeople themselves.[100]

The high school student, Levi Smith, who appeared in the Planet Green series, agreed 100 percent with that assessment. And although he was later to voice criticism of what Planet Green did, Daniel Wallach concurred, in essence, with Kuryak:

I use the argument a lot with both potential donors and some hesitant residents that this kind of exposure and advertising, money could not buy.[101]

Despite their omnipresence, the Planet Green crews were unobtrusive, at least in John Janssen's opinion. He contrasted that favorably with the behavior of networks which came back to Greensburg for the anniversary of the tornado.[102]

Some of the Planet Green crew members simply could not take the destruction that they found on-site, but those who remained were seduced by the austere beauty of Kiowa County and the friendliness of its people.

Two TV crews comprising a total of 16 people commuted daily, six times a week, between Greensburg and Dodge City where Discovery Communications had rented apartments for them. Each crew had, in addition to a camera operator and a director of photography, a sound mixer, an assistant camera and a production assistant. One prominent member of the filming crew was a young woman of Japanese extraction who was known

---

[100] One contract worker related an anecdote to illustrate how comfortable the Discovery crews and townsfolk had become. At a meeting one night in Dodge City, crew members were admonished "by the head honchos" not to fraternize with their subjects.

[101] Gavin Laessig, "Greensburg Returns," *Lawrence.com*, edition of May 27-June 2, 2008.

[102] The CBS program *Early Morning Show* aired each of five consecutive morning shows, from April 27 through May 1, 2008, live from Greensburg.

simply as Junko. Her surname was missing, "just like Cher," she told John Janssen.[103] When it came time for her to return to California, Junko said good-bye personally to each of the townsfolk she had worked with. She told John Janssen that her colleagues in reality TV were jealous of her for having the opportunity to work on a project like the Greensburg series.

****

What did Planet Green show in its three series on Greensburg? The first two episodes of the first series aired on June 15, 2008, and the remaining 11 episodes came out at a rate of one a week. Said Kathryn Takis:

> The focus of 'Greensburg' is to tell the rebuilding of the town with an eye to the future as well as documenting the human struggle of . . . residents putting their lives together.

Planet Green followed approximately 15 storylines. Again, Kathryn Takis:

> The City Administrator, the City Council, the 'green' organizations, the high school kids that are very involved in the rebuilding efforts, the homeowners and business owners that are rebuilding are all stories that we are following.

The network's film crews attended all the public gatherings they could: City Council and business development meetings and school events. "It was a thrill to be able to attend the basketball games as Greensburg made their way to the state finals," according to Takis. "For the May 4 [2008] weekend, we had additional crews attending all the events during that weekend culminating with the visit from President Bush at the graduation."

Takis said that Planet Green had been "on the ground since July so we've been at every event." Filming actually began in late August. Missing the four months immediately following the tornado meant missing a lot of the action. There is a disclaimer at the beginning of the series which alerts viewers to the fact that some scenes were re-enacted for the sake of a coherent story line. One such scene shows Lonnie McCollum telling his wife that everything they owned had been lost in the tornado, so there was no use in staying on in Greensburg. Daniel Wallach found this re-enactment credible. But such an account of the mayor's decision to resign ignores the confrontation with the Estes brothers. This was the clash that precipitated McCollum's on-again off-again decision to quit. That confrontation could

---

[103] A film and TV web site gives her full name as Junko Takeya.

not have been re-enacted easily in the fall of 2007. Ironically this was the flip side of what people had feared—that Discovery would dramatize events and highlight conflict in the series. Perhaps Planet Green was taking seriously what Steve Hewitt had said: The townsfolk would kick the channel out if it showed people fighting.

The Discovery crews, however, did film some conflict in town meetings. Early on, they caught an episode in which one townsman, apparently disgruntled by all the talk about sustainability, is heard to say "I always thought that green was the color of a paint." Daniel Wallach dismissed this as an instance of the series' failure to capture what he called "the multi-dimensional reality" of what was unfolding in Greensburg. "It was not real," he said. "What the man was complaining about had nothing to do with the town's going green." Furthermore, he said, "I know a woman from out-of-state who told me that she was confused by the series' presentation." The Greensburg series also captures remarks by residents critical of the town's leadership. "It's like a dictatorship," one of them grumbled in the episode called *Ice Storm*.

There were more issues than themes, said Kathryn Takis. She went on to list them:

> The cost of rebuilding 'green' and what that means for a small town in rural America. Not only rebuilding, but rebuilding from the ground up. . . . The issue of energy is a major theme. Greensburg is committed to going 100 percent renewable, 100 percent of the time. Again, how does a small midwestern town that was struggling to stay alive pre-tornado redefine themselves as a model for the future? . . . . In a nutshell, it's about rural America coming into the 21st century with a new identity.

In a community as religious as Greensburg—seven churches for a population of 1400—stewardship of the earth provided impetus to going green. The concept came up in a conversation with Bob Dixson. This was the man who defeated John Janssen to become mayor of Greensburg on April 1, 2008—April Fool's Day, as Janssen noted with a hearty chuckle.

"We are not hicks," Bob Dixson said one month after his election. He went on to add:

> We feel called to be here. People in more cosmopolitan settings have a tendency to lose contact with their roots. The tornado was a wake-up call, and we are blessed to have this opportunity [to rebuild sustainably].

Daniel Wallach held a similar view:

> I think the greatest shock to the media . . . is how media savvy these people are. There are stereotypes of rednecks, of folks who live in rural areas maybe not being as bright, and all of those stereotypes have been completely destroyed. . . . L.A. does not have a sense of community anywhere near what Greensburg, Kansas, has.[104]

Kathryn Takis agreed:

> The residents of Greensburg. . .are hard-working people with a strong sense of family values, a deep faith and a sense of good citizenship. These values are what have been the guiding force in getting this community back on their feet after such a devastating event.

Humor was not the series' strong suit. Occasionally its treatment slips into a kind of sentimentality when people are shown struggling to rebuild their lives. Making do in FEMA trailers (to cite one example) is portrayed in a cutesy vignette: A mother is shown making pancakes for her son. You don't hear the smoke detectors going off.

A common complaint expressed in Greensburg about the series was that it missed the emergency phase of the tornado story, but it still evoked sore memories for some. "It was tough to watch," admitted Steve Hewitt, who went on to say:

> At the same time, it gave a reason to bring people together. We are Midwesterners, and we care how we are perceived.

Daniel Wallach conceded that the series offered Greensburg "good exposure," but he criticized Planet Green for a lack of substance and a failure to realize the full potential of the Greensburg recovery story. He quoted Mayor Bob Dixson who said that the series "left the heart of Greensburg on the cutting room floor." For former mayor, John Janssen, the series contained good stuff. He complimented the Planet Green effort as "a reasonable job of what was going on," but he admitted that his wife had not watched the series. Why not? "She said that she had enough reality in her life," he replied with a volley of percussive laughter.

The most acerbic criticism of Planet Green's series on Greensburg came from Gavon Laessig. Laessig had studied television at the University of

---

[104]Gavon Laessig, "Greensburg Returns," *Lawrence.com*, May 27-June 2, 2008.

Kansas. He felt that the series provided political cover for environmentalism: The series gets street cred with Middle America when Greensburg residents are shown to be in agreement with a man like Al Gore, who was always vulnerable to attack as an elitist. Laessig agreed that conflict had been glossed over—to keep Steve Hewitt from growling at Planet Green perhaps?—and situations were streamlined to fit the narrative demands of each episode. Hewitt himself comes across as "an alpha male/reluctant warrior" while the high school students in the series were too idealistic to be representative of their contemporaries in a small Great Plains town. Laessig found the series schmaltzy, succeeding neither as news nor as entertainment. To the series' credit, however, opponents of going green were ferreted out and featured in various episodes.

One observer of the series commented that the second series of six episodes differed markedly from the thirteen episodes of the first season. It was the difference between reality TV—with its stress on characters' reactions to each other and to events—and a documentary series in which the people involved in the filming process understand that they are there to record what is going on and not to intervene in events. Was there a conscious shift in moving from Season One to Season Two? Kathryn Takis denied that there was much difference between the two seasons. Although she wasn't involved in Season Two, she had watched it. In an e-mail (dated March 2, 2010) she said of Season Two that:

> . . .the teams would fly in and out of Kansas documenting the town's progress which just takes time whereas it was important for us to document the initial reaction to the tornado and the initial steps toward rebuilding.

Another TV show about Greensburg was Danny Forster's 4-part miniseries, *Build It Bigger: Rebuilding Greensburg* which began airing Sunday, November 16, 2008, on the Science Channel, another Discovery property. The first episode dealt with the 5.4.7 Arts Center, a project to be covered in the next chapter. Computer animation was brought to bear on some features of the building. The second episode, which aired immediately after the Art Center segment, concerned the Farrell and Debby Allison house, the third with the green rebuilding of the John Deere dealership, and the fourth with a federal subsidy program to help new businesses in Greensburg with their overhead costs. The Allisons were compensated for their cooperation and time commitment in the form of sheet rock and Bosch kitchen appliances. Despite assurances that they would stay out of the couple's way, the film crew did eat away at the Allisons' weekends and evenings, the only time they had to work on the house.

****

Besides national coverage—online for publications like the *Smithsonian* and the *National Geographic*, in standard newspaper articles for others—Greensburg stories began to appear abroad. An example is given by two US-based French free-lancers, Thomas Risch and Sebastien Koegler. Former classmates at a movie school in Strasbourg, France, they went out to Greensburg to interview residents and to see for themselves what was happening. In an e-mail message (dated March 26, 2009), Risch wrote:

> I think that this is a good story. Why? Because we have never seen an example like this [in Europe]! A city, then the tornado, and nothing [left of the city], and the decision to rebuild differently. That's a great story! And the producer, Leonardo DiCaprio, felt this!

A government-funded TV channel, France 24, which offers a French perspective on world news, ran a three-minute segment on what Risch and Koegler produced. Risch felt that Greensburg had the potential for a longer story. He tried pitching it to other French channels without success.

While they were out in Greensburg, the two reporters toured Ron Shank's GM distributorship. Afterwards they ate their lunch in front of the garage. That put them squarely on US 54 which normally carries a heavy load of cattle trucks and long-distance haulers. Not a lunch to compare with a sandwich and a glass of wine on the *Boul' Mich* in Paris' Latin Quarter. Still, Thomas "learned a lot from this lunch." Some passers-by told him that they were not in sympathy with the notion of "this green future." Risch and Koegler wanted to pursue this dissenting view, but they had to rush off to meet Mayor Bob Dixson.

A steady stream of foreign journalists came to town to report on the gospel of sustainable recovery. Wherever news was available, foreigners were reading about Greensburg and watching its recovery on their TVs. Who would have predicted on May 4, 2007, that this small Midwest burg would be world-famous two years later?

THE FRUITS OF GREEN LABOR

Even with eyes protected by the green spectacles Dorothy and her friends were at first dazzled by the brilliancy of the wonderful City. The streets were lined with beautiful houses all built of green marble and studded everywhere with sparkling emeralds.

–Wonderful Wizard of Oz

The principal universities in Kansas are Kansas State University (K State), in Manhattan, and the University of Kansas (KU), in Lawrence. The two are fierce if friendly rivals. Both got involved in the reconstruction of Greensburg.

Twenty-three students from K State worked under their professors' supervision to produce plans for Greensburg which would meet challenges posed in the long-term community recovery plan prepared by FEMA and released in August 2007. Begun in mid-October 2007, the students' work yielded a series of computer renderings of streetscapes, plans, and elevations for the town center. The students' CAD-generated images were to be exhibited at Pratt Community College in December 2007. Nature dictated otherwise: In the first of the hard-luck incidents to bedevil K State work in Greensburg, an ice storm, which cut the college's electricity supply, put the cabosh on that event. The opening was rescheduled, but a big slice of its intended audience never saw the students' work.

Interestingly enough, one of the students' ideas—a 109-foot-tall water tower standing in juxtaposition to the 109-foot deep Big Well in a tourist complex—was similar to what Greensburg actually did to replace its old water tower. An ellipsoid rising 140 feet above the Kansas plain on a site adjacent to the Big Well and visible for miles around was built by

Professional Engineering Consultants (PEC). It was inaugurated on May 3, 2008, one day before the anniversary of the tornado. Described by gun dealer Aaron Einsel as "a ball on steroids," the new tower had a capacity of 100,000 gallons, twice the capacity of the old tower. On the Great Plains, water towers are big deals. "We recognized this new water [tower] was going to be a key symbol for the recovery of this community," Chuck Banks said. "And getting this thing up and operating within one year, before this anniversary, was a high goal of this agency [the USDA]."[105] Funding for the tower, which cost nearly $700,000, came in part from FEMA which contributed $260,000; USDA which gave a grant of $170,000 leveraged by a city contribution of $57,000; and $112,000 from insurance proceeds. "The identity of every community you drive through is somewhat set by its water tower," Steve Hewitt said. "You drive by and see ours, which is now one of the bigger water towers out here."

Another projected change in the physiognomy of the town which influenced K State students' thinking about the remake of Greensburg was a proposed rerouting of US 54. Although some of the students criticized the decision to relocate the route several blocks north of Kansas Avenue, the K State computer renderings which featured views north from that street all took into account the elevation of the new highway. In 2002, KDOT had recommended a new interstate-like road to carry US 54 south of Greensburg. Early in 2008, however, city officials approached KDOT with the idea of relocating US 54, about three blocks north of Kansas Avenue. This re-siting, the city contended, would offer better visibility and access to the town.

In contrast to K State's bad luck, the KU road to Greensburg was a path to glory. The university's project in Greensburg was incontestably the most significant contribution to date in the town's going green. This project was rooted in a graduate-level architecture course at KU known as Studio 804. Established in 1995 by Professor Dan Rockhill, this studio is a design/build course. It demands so much time of its students that they are not allowed to take anything else while they are enrolled in the studio. From its inception, the studio focused on sustainable design. The previous projects undertaken by Studio 804 were private dwellings, located either in Lawrence or in near-by Kansas City. But that changed in a dramatic move, spearheaded by students of the class in 2007-2008.

It was a setback which precipitated this ambitious undertaking. Land in Strawberry Hill, a Hispanic neighborhood of Kansas City where Studio 804 had planned to erect its next house, was sold. So Rockhill's students needed to find something else to do. Said Rockhill:

---

[105] Kevin McClintock, "Freshly painted, water tower is christened," *KCS*, May 7, 2008.

The students do everything. They work with clients, money, finding a site, city officials, manufacturers, supplies, plumbing, heating, AC, using a table saw, laying sod, cleaning up the building for an open house. So many other things are involved, it makes the task of building a relatively minor component.[106]

What his students accomplished in the fall of 2007 abundantly illustrated that remark. Casting about for a new project to work on, Boyd Johnson and Josh Somes, two of Rockhill's students, drove out to Greensburg in October 2007. There they met with Daniel Wallach. They wanted to explore the chances of Studio 804 participating in the green rebuilding of the town. Wallach introduced the students to Steve Hewitt. The concept of a flexible-use building began to take shape in the course of their conversations. The students returned to Lawrence on a high. In Rockhill's absence, they convened a meeting of the students in Studio 804 to share their impressions of what was happening in Greensburg and the idea of designing and constructing a flexible-use building for the town. The students' reactions were all over the place. When he was brought into the picture, Rockhill's reaction was "not negative." "He [Rockhill] leaves no stone unturned, and we had a new stone," Josh Somes commented.

The first that the citizenry of Greensburg got wind of what might happen came on an evening in December 2007, when all 23 students in the class assembled with them to hear Boyd make a presentation before the City Council. Like Daniel Wallach a native of Colorado, Boyd Johnson came armed with computer renderings and a Power Point presentation. What he proposed was mind-blowingly bodacious: A project to be completed in time for the anniversary of the tornado, less than five months away. With Stacy Barnes, Executive Secretary of the Art Council of Greensburg in the audience, Johnson unveiled the concept of a building unlike anything the town had ever seen before. What evolved from his presentation was the notion that this building would be an arts center, with approximately 1,600 square feet of space, to be built to LEED platinum certification. It would be the first such building, not only in Greensburg, but in the whole of Kansas. In fact, there were only 67 such buildings, at the time, in the entire country. Yet, platinum was the standard which the City Council had adopted for all its public buildings with a footprint in excess of 4,000 square feet. None of these buildings had gone up yet, and here was a KU graduate student inviting the city to agree to an arts center, of all things—an arts center, which was a low priority on the list of buildings suggested by FEMA in its long-term community recovery plan—to be constructed to the same demanding standard as the major public buildings of Greensburg. Even in hindsight, the mission seemed quixotic. Its acceptability in December 2007,

---

[106] Charles Higginson, "From the ground up," *KU Giving*, Spring, 2008.

owed much to the heady, sky-is-the-limit ethos of the times. The town's response was positive but "there were lots of holes in the web" to be worked out.

Because Rockhill's students work with modular structure, they can build what they design in one location and move it by truck to its final destination. This was the case with the Greensburg arts building. Its modules were built in Lawrence, a few miles from KU. The studio's building came to be known as the 5.4.7 Arts Center. The numbers were chosen to commemorate the date of the tornado, which occurred on the fourth day of the fifth month of 2007.

"I think we've lost our way in the last 30 or 40 years with the sort of building boom and the enthusiasm for suburbia, and we have lost sight of the sensible way to build," said Rockhill. He is probably the best known professor of architecture at KU.[107] With an interest in inexpensive housing, Studio 804 has garnered several Affordable Housing Awards for its private dwellings in Kansas City and Lawrence. The studio course has also reaped a slew of additional awards and nation-wide publicity for Rockhill and his students.

Interviewed for a profile which appears online in *AIArchitect This Week*, Rockhill said:

> We've done the last four houses [2004 through 2007] as prefab projects, meaning we prefabricated in Lawrence and then trucked it to Kansas City. . . . So what we did was build a really modern house in Lawrence, trucked it to Kansas City, and put it on a site that nobody believed they'd put a house on. It sold right away. . . . The neighborhood started fixing itself up, which was a benefit, so the next year we went to a different neighborhood. [Again] the house sold right away. Then the same thing happened with the third house.
>
> . . .
>
> Where I live, modern design is seen like the bird flu. . . . So wherever we go, we're always criticized because the work stands out. It completely confounds our critics when these houses sell instantly and nobody is interested in the houses that they're struggling to sell for half the money.[108]

---
[107] Brendan Lynch, "Remaking Greensburg," *www.researchmatters.ku.edu*, February 2008
[108] Heather Livingston, "Dan Rockhill," *www.aia.org*, August 3, 2007.

Rockhill works with community development corporations. His last two houses in Kansas City, Kansas, were done with *El Centro*, an organization that works exclusively with the Hispanic community. *El Centro* underwrote both projects, and Rockhill returned it a percentage of his profit. He gets no financial support from KU although he works for the university. Studio 804 is a completely independent not-for-profit corporation. If there is money left over at the end of a project, he may return some of it to an underwriter like *El Centro*. Whatever money he earns is absorbed by the business for such things as the purchase of tools.

In November 2007, a couple of students from Studio 804 attended, at their own expense, the USGBC exposition in Chicago. This was the show where Planet Green had filmed John Picard in conversation with Greensburg high school students. Their aim was to drum up support from vendors and donors for the structure the class was in the process of designing and would soon be building. The client for the project was not the city of Greensburg itself, but the Greensburg Art Council, a private, non-profit organization. Its executive director, Stacy Barnes, had lived for a while in Lawrence before returning to her home town of Greensburg. As soon as she was back, she went to work as an assistant to Steve Hewitt. Stacy met Dan Rockhill after Boyd Johnson's presentation before the City Council. She gave voice to her vision of a new arts center at that time. This piqued Rockhill's interest, but there was not much time for Barnes and Rockhill to interact. Nonetheless the project got the go-ahead almost immediately. Financial arrangements between the Art Council, with its 8-member board of directors, and Studio 804 were pretty loosey-goosey. Since the studio is itself a not-for-profit organization, its funding comes in part from third-party contributions. Stacy Barnes knew that she would have to hand over a check for $10,000 before the building could be moved out of Lawrence, but her understanding of the amount owed KU was of the order of $200,000-$250,000. The exact amount was determined only after the building had been transported to Greensburg, reassembled, and the finishing completed. And that amount turned out to be $325,000, a lot higher than the Art Council had anticipated paying but considerably less than the $450,000 which FEMA's Long Term Community Recovery plan had projected. Still, there was some grumbling. In an interview which appeared on August 1, 2008, in the *Signal*, John Janssen vented some spleen on the subject:

> The glass house is a deal that got brought into the community and then sucked up $500,000 (*sic*) in resources that could have been used elsewhere. . . . They came to us last winter and told us they needed $75,000 to come to town with one of their modular unit buildings and we said we didn't have it. . . . Next thing we know the trucks are coming to town with the cameras rolling. They dump the boxes here and say, 'By the way, it's

now half a million we need.' . . . KU wanted that sucker in town and they got what they wanted.[109]

More than three months later, Aaron Barnhart of the *Kansas City Star* weighed in with his own comment about the Arts Center and its cost. Interviewing Bob Dixson, Barnhart noted that the Art Council director, Stacy Barnes, was the mayor's daughter and went on to write:

> The city still needs to find $350,000 to pay for the 5.4.7 Arts Center. In the redevelopment parlance of Greensburg, this deficit is known as a "gap." I asked Dixson about it. He looked me square in the eye and said, 'The gap will be met.' He looked at me for a moment, then added, 'That's it.'[110]

****

The facility which Rockhill used for several years to manufacture his prefabricated houses lay in an abandoned industrial zone just to the east of Lawrence. Formerly occupied by Farmland Industries, which produced fertilizer there until 2001, the site comprises 467 acres of what became a nitrate-saturated wasteland. The city of Lawrence long played with the idea of buying the property in order to rehabilitate it as an employment center and community open space. (It lies just to the west of an existing business park.) At first, offers to buy the land foundered on the issue of clean-up. Lawrence insisted that the property's trustees assume responsibility for the site clean-up: This could have taken as long as 30 years to complete. Amidst a jumble of rusting pipes and storage tanks stood a warehouse where Studio 804 students labored under draconian conditions to construct the Arts Center modules.

On March 5, 2008, the students in Studio 804 were bustling about, dressed in parkas, heavy work gloves and caps with ear flaps. They looked as if they were getting ready for an Iditarod. The temperature in the warehouse where they were asembling the seven modules of the Arts Center felt to be below freezing. "We can judge how cold it is in here by the number of overhead light bulbs that are out," a student guide said. On March 5, only one light bulb was out of action. One could only imagine what it was like when all the light bulbs failed. (This occasionally happened.) In one

---

[109] Mark Anderson, "Though out of office, Janssen still has plenty to say," *KCS*, Aug. 1, 2008.
[110] Aaron Barnhart, "The greening of Greensburg continues this week on TV," *blogs.kansascity.com*, Nov. 8, 2008. The *KCS* reported on Feb. 17, 2010, that, through a combination of private donations, a matching grant from the South Central Community Foundation, and fund-raising at KU, the Art Council had paid off what it owed on the facility.

corner of the hangar-like building, Planet Green had attached to the rafters a time-lapse camera which took periodic snapshots of the floor below.

The goal that Studio 804 had set for itself was a *tour de force*, a project, done from start to finish, in less than 20 weeks. That included the transport of the seven individual modules over 400 miles, the longest distance by far that any of Rockhill's structures had ever travelled.[111] Furthermore, the Arts Center was to be built to meet the criteria for LEED platinum certification, the first time that the studio had attempted to do so. This was a tall order by anybody's standards.

The process began in January 2008. First step in the construction phase of the project involved a demolition job: Contacted early in 2008 by Sunflower Redevelopment which controlled a decommissioned ammunition plant east of Lawrence, Studio 804 was invited to salvage lumber from a magazine building. This was one of approximately 1,000 buildings on the 9,000-acre site. The Sunflower Army Ammunition Plant had been established in 1941. It provided munitions during three conflicts: World War II, and the wars in Korea and Vietnam. The use of recycled material has been a staple of Studio 804 from the beginning. Rockhill seized the opportunity for his students to make off with inch-thick boards of Douglas fir, cedar and pine. The Douglas fir was destined to sheath the north and south sides of the Arts Center box. The cedar was to skin the east and west sides of the building. Students spent five days in January stripping 10,000 square feet of wood off the magazine shed and hauling it back to the warehouse in Lawrence. On each of those days, the students were escorted to the magazine building by Sunflower guards. The weather refused to cooperate—a constant refrain in much of the recovery and reconstruction work undertaken in Greensburg. It rained and it snowed. Sleet pummelled the students as they worked. The boards they ripped off their target shed flew out of their raw, red hands.

In mid-March, 2008, there was a strategy session in Rockhill's warehouse. His students needed to do a last-minute checkout after their seven modules had been winched onto seven flat-bed trucks. The weather was no milder than it had been earlier that month. An overtaxed heater warmed the room where the meeting was held. The rest of the warehouse was deprived of even this Scrooge-like luxury. Parkas stayed on, however. Only some of the work gloves and caps came off. One of the students referred to this conference room as "the sauna." Pushed to the side of the room was a table bearing an assortment of goodies—doughnuts, bananas, strawberries, and coffee—nourishment to keep the students going. They sat in an oval, with Dan Rockhill located at one end, his jeans-clad legs spread out in front of him and crossed at the ankle. Rockhill is a tall man with white hair

---

[111]The distance from Lawrence to Greensburg is 270 miles, but the size of the modules made it necessary to choose a roundabout route to avoid obstacles like overpasses.

combed straight back. His is a commanding presence. Although students address him by his first name, familiarity does not mask their respect for the man.

The discussion of last-minute details was astonishing for the give-and-take among the students and between them and Rockhill. True, as the architect has asserted, the students take it upon themselves to make crucial decisions, voting if need be among themselves, always with a porous mix of candor and passion, on what to do and where to go. Studio 804 is structured around a division of labor. Each student is assigned a sub-task. Debriefings are the way students share information with their comrades. This particular day, a young woman spoke about the furniture possibilities she had researched. The meeting adjourned after a couple of hours.

Students, most of whom had been putting in 18-hour days, needed to come up for air—"time to do laundry and say good-bye to our loved ones"—before the trek across the prairie. The modules of the Arts Center had already been disassembled and hoisted onto the semitrailers which would crawl down a network of secondary roads to the site opposite Greensburg's Big Well complex and new water tower. This would be the building's final resting place. After some debate, Rockhill's students settled on a compromise time of 6:30 a.m. to meet the following day.

The weather on March 16 could not have been much worse—"the worst rainstorm ever" in Rockhill's estimation. The truck convoy and a car carrying three students encountered what the *Lawrence Journal-World* described as "driving wind and pounding rain" on its eight-hour journey from the construction site at Farmland Industries to Greensburg. At one point, a student braved the downpour to clamber up one of the modules to secure a weather-proof membrane which had been torn loose by the wind. Further down the road, students had to retrieve some of the temporary siding which had blown off another of the modules. "We had a couple of loose pieces," Rockhill said, "but considering the thousands of pieces we used, that wasn't bad."

Over the next couple of days, studio students planed off the foundation walls and craned the seven modules into place. Planet Green was there to film every iota, shooing away other photographers whose presence undercut the impression that only Planet Green was on site to record the event.

Rockhill's students spent the next two months, from mid-March to mid-May, working in Greensburg. They lived at Pratt Community College, where the ill-fated K State exhibition had been booked. They commuted the thirty miles to Greensburg every day, arriving at dawn and staying till nightfall. A daily buffet lunch was prepared for them and served in a garage close to their work site. Rockhill himself stayed at what one of his

Unloading one of the Arts Center modules

students described as a second-rate hotel. He worked almost as many days in Greensburg as his students.

Since the students were tied down to their work site for the whole day, they made do for snacks with what the Qwik Shop two blocks away could provide. Theirs was a daily diet of sandwiches, sausage rolls, and soft drinks. After a while, it was easy to spot Studio 804 students. Their faces had the color of boiled lobsters, brought on by wind burn. Working conditions were different from what they had been in Lawrence but just as challenging. Slogging through mud took getting used to. Remember, they were told, "clay is slicker than tiger shit." Camera crews from Planet Green swarmed like mosquitoes and became, in the words of one of the students, "part of the tapestry."

The KU students referred to their building as a sustainable prototype. Rockhill's opinion was that "this could conceivably be something that leads to a whole lot of material . . . to rebuild Greensburg."[112]

To win LEED platinum certification—a process which, according to Rockhill, makes "doing your income tax look easy"—the 5.4.7 Arts Center involved, among other features, the following:

---

[112] Jonathan Kealing, "Architecture students to demonstrate how Greensburg can rebuild greener," *Lawrence Journal-World*, Dec. 5, 2007.

- Cellular fiber insulation from recycled newspaper
- Three wind turbines to generate electricity
- Three geothermal wells to take advantage of the earth's near-constant temperature
- Photovoltaic roof panels to utilize solar energy
- Sedum planted on the roof as a garden to moderate extreme summer and winter temperatures
- A system to capture and store rainwater in a 1,500-gallon underground cistern
- A green glass "skin" covering the walls to protect the recycled wood stripped off the shed at the Sunflower Ammunition Plant from ultraviolet sun rays.

More than half the south façade was taken up by a wall of glass doors which permitted large art work to be passed in and out of the building. The wall was protected by a hinged lift-up—normally used in aircraft hangars—a one-piece hydraulic door manufactured by HydroSwing. A long panel of louvers offered protection against the sun as well as adding an esthetic complement to the straight-line geometry of the Arts Center's design.

The 5.4.7 Arts Center was not totally finished by the time of the tornado's anniversary, but it was far enough along for Dan Rockhill to receive visitors inside the building on May 4, 2008.

"We work at warp speed," Rockhill said. "People cannot understand how hard it is to construct a building in the time we do."[113]

When it was announced that the 5.4.7 Center had achieved platinum certification, praise rolled in. Said Rick Fedrizzi of the USGBC:

> Studio 804 is to be congratulated for receiving platinum LEED certification. The Arts Center will be a showcase for high-performance, energy-efficient, healthy buildings, and an inspiration for others.[114]

KU *did* get a lot of mileage out of the project. It was featured prominently in the spring 2008 edition of *KU Giving*, a publication of the fundraising foundation for the university. The magazine cover showed one of

---

[113] *Ibid.*

[114] J-W Staff Reports, "KU-made building certified platinum for being green," *Lawrence Journal-World*, June 7, 2008.

The 5.4.7 Arts Center

the studio students, hammer in hand, and dressed to work in the warehouse at Farmland Industries. The same issue listed a number of contributions to Studio 804, including one from Google for $100,000. This particular gift raised hackles in Greensburg. A couple of residents privately expressed concern over the use by KU of the 5.4.7 Arts Center to raise money for the university. Implicit in all this, there was a sense of exploitation although that word was never actually used.

One of the studio students, Lindsey Evans, of Belleville, Illinois, said:

> I've learned more in the last four months than in the last five years. Something comes from experimenting with materials—cutting the wood, seeing the grain, holding the concrete. Knowing what the materials can do allows you to make the best use of the materials.[115]

For his part, Rockhill stressed how the 5.4.7 Arts Center could serve as an example for other environmentally sensitive buildings:

> Greensburg will hopefully see Studio 804 as providing . . . an example that, if students can do it we can certainly do it—meaning both the community as well as people that they bring into town to do development.[116]

---

[115] Charles Higginson, "From the ground up," *KU Giving*, Spring, 2008.

[116] Brendan Lynch, "Remaking Greensburg," *www.researchmatters.ku.edu*, Feb. 11, 2008.

The Arts Center attained iconic status as soon as it opened. Besides the exposure in the spring 2008 issue of *KU Giving*, articles about the Arts Center appeared in prestigious national magazines like *Metropolis* and *Architectural Record*.

****

In the spring of 2008, Greensburg GreenTown moved to a small frame house located at the corner of Sycamore and Florida Streets. This was the site where Lonnie and Juli McCollum's dream house had stood. Jerry Diemart, who had joined the Board of Directors of Greensburg GreenTown, proudly showed off what he described as "an inverted DC solar heat pump." Powered by a wind generator and two solar panels, the unit used 50 percent less energy than comparable units on the market. Daniel Wallach had secured it as a donation from an Arizona-based company.

By the summer of 2008, Greensburg GreenTown had developed a division of labor. It hired two recent graduates from Washington University in St. Louis, Emily Schlickman and Mason Earles, to spearhead the fund drive for the chain of eco-friendly houses, an ambitious endeavor to get two million contributions of $5 each. This was ultimately abandoned in favor of a scaled-down project to build houses as money for them became available.

Greensburg GreenTown had put the students up in a bungalow on the east side of town. A jumble of bicycles was stacked along the rear wall of their office-home. These were to be painted lime-green and put to use as an alternative form of locomotion. Donated by Sunflower Resource Conservation and Development (RCD) and refurbished by prison inmates at a Kansas correctional institution, the bikes were supposed to be offered free of charge to townsfolk and visitors alike the following year.[117] The idea was to copy programs, like *Vélolib'* in Paris, where residents pick up bikes at any of many locations, peddle about to their heart's content, then leave them off at one of the bike stands in the central city.[118] The Greensburg program would not have had quite that much flexibility, but it would still

---

[117]This program adopted part of the urban philosophy of BNIM. People would foresake their pickups and cycle around town, in an environmentally-friendly fashion. This dodo, alas, would never fly. Residents were not about to ditch their four-wheel drives to run errands on two wheels. In the fall of 2010, a peek through a window of the house that had housed the offices of Greensburg GreenTown in 2008-2009 revealed a stack of bikes ready to be shipped out. It was the bikes, not the pickups, that were given up.

[118]This idea has been tried elsewhere in the US. A company called Alta implemented a singularly successful program in Portland, OR, in the 1990s. Alta is a 17-city organization which was prepared to launch, during the summer of 2012, its biggest ever bike-sharing program in New York City. See the profile of Alta president, Mia Birk, in the January 2012 issue of *Metropolis*.

have offered an alternative to driving a standard vehicle.[119] At about the same time, Greensburg GreenTown began to organize one- and two-hour tours of green structures in the city, charging $5 per person per hour.

As students, both Mason and Emily had been interested in green building and sustainable design. Emily majored in environmental and international studies at Washington University, while Mason did urban studies. They toyed with the idea of doing an independent project, possibly in Oklahoma—Mason was born and grew up there—bringing to his small hometown green builders from Oklahoma City and developing "green collar jobs." Clearly this was a creditable goal, but probably too much for freshly-minted undergrads. Then Emily's mother saw a TV program about Greensburg. She did some Internet research on the town and put the two young students on to the idea of working for Greensburg GreenTown. Emily and Mason began to correspond with Daniel Wallach. In the end, the three of them put together a grant proposal for AT&T. When it was funded, the money—$50,000—was dedicated to overhead, including the students' salaries. After their first two months in Greensburg, however, they were able to move under the auspices of AmeriCorps. This was the volunteer-coordinating agency that the Lawrence-based anarchists had briefly worked for one year before. Money from AT&T was shifted over to directly support their project. This was not the first donation which AT&T had made in Greensburg. Previously it had invested $2 million in the town's infrastructure, laying ten miles of new line, as well as plastic duct work to allow for the future installation of optic fiber at public buildings. At the time the company made its contribution to Greensburg GreenTown, AT&T Kansas President, Dan Jacobsen, said: "AT&T is a strong supporter of the responsible use of our natural resources and inspiring innovation, which is exactly what the Chain of Eco-Homes is all about—incorporating new technologies and sustainable design into the rebuilding of this community."[120]

What exactly was a chain of Eco-Homes? They would act, Emily said, as informational centers for visitors and residents interested in state-of-the-art green building techniques like ICF (Insulated Concrete Forms) or SIP (Structural Insulated Panel) construction, or how a radiant floor system works. (SIP was the construction method employed by the Mennonite builder, Lloyd Goossen.) It would be possible for visitors to stay overnight

---

[119] Road conditions, as late as the end of 2009, turned driving—much less cycling—into a teeth-jarring adventure. Potholes and gulley-like depressions at street intersections made getting about town a bit like negotiating an obstacle course. Repair work in the fall of 2009 left runnels between recently installed curbs and the old street pavement. These troughs were just about the width of a car tire and a couple of inches deep. Although marked off by orange-and-white traffic cones, these tire traps snared incautious prey. One such carcass, a terminally ill Chevrolet, was on view that year at Shanks Motors garage.

[120] "AT&T Delivers a Most Generous Gift," *www.greensburggreentown.org*, Dec. 8, 2008.

in what might then be called eco-lodgings. It was planned that the houses would be dispersed throughout the center of Greensburg and linked by something like a bike path. The houses would be different sizes and would involve a variety of systems to show that it was possible to go sustainable in different ways. Emily and Mason described themselves as the glue which held together the various teams required for the eco-homes, the architects, engineers, product suppliers, and manufacturers.

Construction of Greensburg GreenTown's demonstration houses lagged for a variety of reasons. The sequence in which they went up depended on the availability of resources. When *Mother Earth News*, a national publication headquartered in Topeka, found out that its proposed house would have to be handicap-accessible, that raised the price tag from $250,000 to $420,000, and that shoved the *Mother Earth* house to the back burner if not entirely off the stove.[121]

Another of the Eco-Homes to lack funding was a house designed by a Kansas City-based architect, Robert McLaughlin. Emily and Mason were working with the Kansas City chapter of the USGBC to get LEED platinum certification for this 1400-square-foot, ICF structure to be called Greensburg House X.

The last of the no-shows/yet-to-appear was a SIP-construction building, designed by K State University students working under Prof. Gary Coates, in partnership with a California-based company. The company, then called Xtreme Homes, had agreed to undertake the funding for the project. This pre-fabricated, modular structure, like the 5.4.7 Arts Center, was to be built off-site, in this case in the California factory of Xtreme Homes. Again like the Arts Center, it would be trucked to Greensburg. Under a new name, Xtreme Structures, the company had been aggressively searching for a Midwestern site in a green, sustainable community where it could build a new factory. Greensburg was an obvious candidate for such a site. There was even a rumor that the company would break ground there in the spring of 2009. In order to entice the company to build in Greensburg, the city was willing to issue an IRB (Industrial Revenue Bond). None other than John Picard had put Xtreme Structures in contact with Greensburg officials. Collaboration with the company proved to be a chimera, however. This was another example of the rotten luck K State had in Greensburg. Xtreme Structures went 'xtremely' bankrupt before anything could happen.

As of the spring of 2011, the one success story that Greensburg GreenTown could tout involved a pre-cast concrete home called the Greensburg Silo House. The partner for this project was Armour Houses, a company headquartered in Florida. Armour's interest lay in constructing a building which would withstand violent winds like those of the tornado of May 2007.

---

[121] Four years after the tornado, only one of the Eco-Homes had gone up; a year later, a second was under construction.

Yet one more house was on the drawing board. Its design was being done by students at the University of Colorado School of Architecture in Denver. The students intended to work within strict financial constraints—a budget of $50,000 has been proposed for the building's materials—to produce a straw bale house. Commercially this material was called *agriboard*, "the waste straw from the manufacturing process (that) can be reused as landscape mulch or animal bedding," as explained on the web page of the firm that makes the product, Agriboard Industries, of Electra, Texas. Agriboard was to be Greensburg GreenTown's partner in this endeavor.

In the spring of 2009, the Agriboard plant in Electra went up in flames, victim of a grass fire. The owner of Agriboard, Ron Ryan, formerly of Wichita, told *The Wichita Eagle* that he planned to move his business, lock, stock, and barrel, from Texas to Greensburg. Contact between Ryan, who founded Ryan International Airlines—he sold out in 2004—and Greensburg went back to late 2007 when the possibility of an Agriboard expansion had been explored between Ryan and the Kiowa County Commission. It now appeared that the company would simply relocate to Greensburg. But Ryan seemed not to have consulted Agriboard President Mike Huskey. According to Huskey, relocation was only an option, not a decision already taken. Steve Hewitt was guardedly optimistic about snaring the manufacturer of compressed wheat-straw panels for Kiowa County. Officials from the company had met with him and Bob Dixson in mid-April, 2009, to talk about the incentives which could be offered Agriboard to lure the company into relocating. "We've been working very hard with them," Hewitt said. "It's not a done deal . . . but if we can make an offer appealing enough, they'll come."[122] Not to be, sad to say. The Agriboard move came up a cropper and joined the list, which grew longer with each passing year, of industry-snagging debacles.

After the big-city attractions of St. Louis, did the ex-students of Wash. U. suffer from culture deprivation on the plains of south-central Kansas? Emily said that they were excited about the projects they were promoting. They had logged hours and hours working on them. Both were vegetarians, so the choice of restaurants available to them was severely limited. They shopped for groceries in Pratt and spent what little free time they had in exploring the countryside around Greensburg. Both students said that they were encouraged to see how townsfolk were going green on their own, in a variety of effective but non-sensational ways. They cited as an example what Rex Butler had experimented with: attaching his attic ceiling fan to an alternator to generate electricity.

---

[122] Mark Anderson, "Greensburg hoping to lure Agriboard's 42 jobs to town," *KCS*, June 5, 2009.

Emily's and Mason's commitment to working in Greensburg was not open-ended. Both of them had plans to apply to graduate school in architecture. They talked of applying to a variety of prestigious institutions which included Harvard and M.I.T.

****

At a January 2008 meeting of the City Council, hospital administrator Mary Sweet reported that environmental and soil studies of the new hospital's site—where the BTI John Deere facility had stood on Kansas Avenue—had been completed. The architects for the new building, Wichita-based Health Facilities Group, were consulting with BNIM and NREL to design a hospital which would meet the criteria for LEED silver certification. But Sweet cautioned that a new standard for hospitals was being drafted by LEED. Plans for the Kiowa County Memorial Hospital might be upgraded to meet the new standards. The price tag for the new hospital was placed at $25 million. That might have been a deterrent, but the need for an on-site facility was perceived to be urgent. Driving to hospitals in towns like Great Bend, 100 miles away, or even Pratt, only 30 miles away, wasted a lot of time.

On the tornado's first anniversary, hospital employees served guests from two mirror-image cakes, one decorated with the ruins of the old hospital, the other with an artist's rendering of the new hospital. The first cake bore the words "From Tragedy" and the second the words "To Triumph."

Ground-breaking for the new Kiowa County Memorial Hospital occurred half a year later, on October 28, 2008. The ceremony took place under a big tent on US 54 across the street from where the new facility was to be built. Trucks roaring down the highway a few feet away repeatedly drowned out the speakers. A stiff wind gusted through the tent. But neither of these distractions could dampen the spirits of the participants. An ebullient Mary Sweet acted as Mistress of Ceremonies. There was as well a host of dignitaries: State Secretary for Health Rod Bremby[123]; representatives of both of Kansas' US Senators; and Lieutenant Colonel Tim Stevens, who had overseen the erection of the EMEDS 18 months before. Representative Dennis McKinney, ever ready with an apposite Biblical quote, reminded his listeners that what they did for the least of their brothers, they did for Jesus.

Although the new hospital would be about the same size as the one it replaced, with one department—behavioral health—moving to Kinsley, it would be a lot greener. Concrete from the former John Deere parking

---

[123]Bremby played a pivotal role in the dispute over energy generation in the State of Kansas, an issue to be explored later.

lot would be recycled for the new building. Some of its own energy needs would come from a wind-powered generator. Mary Sweet's remark to the City Council that the new hospital's environmental standards might be upgraded was prescient. The new hospital would be built, in fact, to meet LEED platinum certification.

Still, there was no pleasing everyone. That included the ex-mayor, John Janssen, who groused that: "We need the hospital, and let me be clear about that, but it needs to be thinking outside the box a little bit more and become a more viable entity in the community." He went on to say:

> [Hospital authorities] have had two different medical groups from Wichita, Galachia Medical Clinic and Via Christi, offer to be a part of building the new hospital . . . But both groups were basically told we weren't interested. [Some of us] met in Wichita with the Via Christi folks and they sat down and said 'Here's what we can offer you.' It looked good . . . but when we got back home the rest didn't want any part of it. To me, Great Plains, which manages Kiowa County Memorial Hospital, is just a glorified accounting firm. They can't offer us what Galachia or Via Christi can.[124]

Marvin George, pastor of the First Baptist Church and member of the hospital board, responded to Janssen's criticisms, and he was just as outspoken as Janssen had been: "John has a responsibility to get the facts straight before [speaking out]."[125] George continued:

> The board did meet with Galachia early this past year . . . They were of the understanding we weren't going to have a hospital and they were going to come in and build a private hospital and put the bill on the county. . . As for Via Christi, in a phone conversation they never had any desire to come here and build or help build a hospital . . . We'd been working in a cooperative manner in transferring patients to them, and that relationship will still exist.
>
> . . . .
>
> Give credit where credit is due. Great Plains Health Alliance has never once backed away from us, or our aim to provide health care for the community.[126]

---

[124] Mark Anderson, "Though out of office, Janssen still has plenty to say," *KCS*, Aug. 1, 2008.
[125] Mark Anderson, "George counters Janssen's take on the new KCMH," *KCS*, Aug. 8, 2008.
[126] *Ibid.*

The squabbling between Janssen and George might not, in itself, have amounted to much more than a pissing contest. But it adumbrated a crumbling of community unity in going green. A convergence of opinion on the rebuilding of public buildings in Greensburg began to unravel.[127]

Early in March 2009, Mary Sweet told Mark Anderson that she hoped to avoid issuing revenue bonds to fund a gap of $5.1 million between the projected budget for the new hospital and funds-in-hand. Negotiations were underway with the USDA to close the gap. Sweet offered an example to illustrate why it would be difficult to meet LEED platinum certification:

> You get points toward certification if you get your materials from within a 500-mile radius. We could only get gray cement for the concrete exterior of the building in that radius while our original plans called for more of a tan/sandy color. To get that kind of color, you need to work with white cement to make the concrete, and we couldn't get that within 500 miles. . . . We wanted a healing, warm appearance. . . So we went with the white cement, even though it cost us some credit on the LEED scale.[128]

Mary Sweet had problems to deal with, other than LEED certification. "I've heard rumors," she told Mark Anderson, "that the project was shut down because of a lack of funding." Moreover, she said, "My construction manager even heard it when he went across the street to the Lunch Box one day; someone told him it was too bad he'd be out of work soon."[129]

Popular support for Greensburg continued to be expressed throughout the state. A poignant example of this was provided by the small town of Baldwin City which lies a short distance south of Lawrence in northeastern Kansas. There, a quilter named Judy Johanning decided to organize a quilting bee for Greensburg. News of the project spread by word of mouth. Ultimately some 40 quilts were made and distributed, with Mary Sweet's help, to employees of the Kiowa County Memorial Hospital. Some of those hospital workers helped make the quilts. "It's getting really cold," Mary said at the time. "There's always a need for blankets."

By ground-breaking time for the new hospital, Chris Gardiner was working only one day a week in Kiowa County. Learning that it would cost $200,000 to replace the house in Greensburg which they had bought for $75,000, Julie and Chris reluctantly decided to move 87 miles away, to

---

[127] This will be exemplified in the disputes which erupted over plans for the Twilight Theatre and the Big Well Museum.

[128] Mark Anderson, "Sweet: new hospital construction chugging along," *KCS*, March 5, 2009.

[129] *Ibid.*

Hutchinson, where Chris got a new job. The family found a house that it could afford. In the fall of 2008, Megan Gardiner enrolled at Pratt Community College in a program that led her to develop an interest in pre-medicine. In light of Chris' profession and her own actions on the night of the tornado, that seemed only natural, even inevitable. After her graduation from Pratt in the spring of 2010, Megan was accepted by the radiology program at Fort Hays State.

The family's pug, Cheyenne, died only one week after the tornado, but the Gardiners' boxer, Joe, was doing well. Julie Gardiner did say, however, that he got skittish every time the weather took a turn for the worse. Thunder and lightning sent the dog under cover, quivering with memories of May 4, 2007. The Gardiners bought another boxer to keep Joe company.

The Gardiners' boxer, Joe

TRAGEDY TO TRIUMPH

"The Silver Shoes" said the Good Witch, "have wonderful powers. . . All you have to do is to knock the heels together three times and command the shoes to carry you wherever you wish to go."

"If that is so," said the child, joyfully, "I will ask them to carry me back to Kansas at once."

–Wonderful Wizard of Oz

Wind power, as both Lynn Billman of NREL and Steve Hardy of BNIM insisted, was an obvious source of energy for the rebuilt town of Greensburg. In fact, there was an announcement at the time of the *second* anniversary of the tornado that a wind farm, capable of generating 12.5 megawatts of electricity, would be built by John Deere southwest of Greensburg. But that is getting ahead of the story.

Wind is an enduring fact of life on the Great Plains. It was wind that had destroyed the small city in 2007. It was wind that would generate power for the rebuilt town. Yet another manifestation of wind threatened to disrupt the tornado's first anniversary celebration. The large white tent erected on Thursday, May 1, 2008, to accommodate a number of activities, including a picnic supper the following Saturday and a communal religious service on Sunday, collapsed under the pummeling of winds that gusted up to 50 mph. The storm passed through Greensburg on Friday, May 2, 2008. It went on to tear off roofs and knock over train cars near Kansas City. More than 350 houses were destroyed. Building projects like the 5.4.7 Arts Center were disrupted. People working outside, like the Studio 804 students, wore sunglasses and overcoats and walked backwards to protect themselves from the airborne grit. The winds did not die down until Saturday morning when the tent in Davis Park was re-erected in time for that night's events.

Earlier in the day there had been a display of antique cars along Main Street. Lonnie McCollum would not normally have passed up such an event, but his 1932 Ford pickup had been badly damaged by the storm. At first, he did nothing to restore it. Friends persuaded him to reconsider. The pickup was put back together by a friend in Bucklin. Nonetheless McCollum did not return for the anniversary of the tornado which had altered the course of his life so much. "I have no plans to be there this weekend," he told Mark Anderson of the *Signal* to whom he confided that: "I haven't been a part of [the rebuilding], but I'm encouraged by what I see developing."[130]

Life had not been easy for the McCollums after the tornado. They rebuilt just outside Pratt, 30 miles to the east of Greensburg, where McCollum took a full-time job as a building inspector to help pay their mortgage. But the new house had serious faults which the McCollums felt they were not up to fixing. The house went on the market in the spring of 2008. It had to be sold at a loss. "Lonnie tells people that he is desperately searching for the life that he lost," one of his friends said. "I had a Norman Rockwell existence in a great town with a nice property, free of debt, in an ideal setting," McCollum told Mark Anderson.[131] The former mayor of Greensburg and his wife found some solace in a new church they joined. They bought yet another house in Kinsley after their house in Pratt sold.

The day before the anniversary, ground was broken on the first of the major civic projects to be constructed. This was the business incubator, to be built to LEED platinum certification at a price tag of $3.5 million. This was the first of the major civic buildings to go up in Greensburg. The ribbon-cutting ceremony for the incubator occurred a year later.

The theme for the anniversary was *Tragedy to Triumph*. Greensburg planned three days of events to celebrate the rebirth of the town. Sunday, May 4, 2008 began with a joint religious service in Davis Park. So many people showed up, there wasn't enough room for everyone in the re-erected white tent. Marvin George, pastor of the Baptist Church—a church which was supposed to be constructed to the LEED specs for platinum certification — presided. Several other Protestant ministers participated. All their churches had been flattened by the tornado. Representing the Mennonite Church, Pastor Jeff Blackburn recognized the 18 members of the graduating class, including Megan Gardiner. Their commencement was scheduled for that afternoon. Blackburn told his audience:

---

[130]Mark Anderson, "Lessons aplenty in response to disaster," *KCS*, May 4, 2008.
[131]*Ibid*.

> These students have often flown under the radar. You know, people spend their entire lives thinking about themselves, demanding their own way or throwing a fit to get their own way. And yet this class has realized a long time ago that it's not all about them.[132]

Pastor Marvin George, who had served in the US Army before going to seminary, revved up the congregation by repeating the theme of the day, Tragedy to Triumph, getting everyone to join in on the word 'Triumph' like a football cheer. "I've been asked what's the difference between this time last year and this time this year," George said.

> Well, Lord have mercy, triumph—that is the difference. We ought to be excited. Are you excited to be here? Then smile! It increases your face value. . . . I am glad I'm alive and well at this time and at this place that God has ordained for us to triumph. And to tell the rest of the world that we are triumphant. And they are watching. They're watching.[133]

Thanks to Planet Green and film crews from several national and local channels, the rest of the world was certainly watching. On the perimeter of the tent, a phalanx of cameras had staked out their ground, one beside another, like the cavalry in a Western, ready to charge, ready to record all the singing, all the yelling, all the words of comfort that reverberated across Davis Park that morning, from 10 a.m. to noon.

Scattered throughout the audience were men and women dressed in leather chaps and jackets with slogans like "Soul Seekers for Jesus" and "Patriot Guard Rider." To honor the 2008 graduates of Greensburg High School, bikers from across the state of Kansas converged on the town by the hundreds. One observer put the number of bikers at 1,000. The Patriot Guard maintained a high profile throughout the day.

The joint service in Davis Park disbanded to allow people to queue up for a picnic lunch of chili dogs and potato salad. While Planet Green interviewed Marvin George, the Patriot Riders made their way back to their Harley-Davidson hogs. After half an hour of milling about, they roared off, *en masse*, down US 54 to parade around the streets of the town, some of which were near-empty. Overhead, helicopters crisscrossed the sky above Greensburg, drowning out conversations on the ground below. They were there as part of President Bush's return to Greensburg almost a year after his first visit to the town.

---

[132] Kevin McClintock, "Community pastors share message of hope in Davis Park," *KCS*, May 7, 2008.
[133] *Ibid.*

Patriot Riders

There were other out-of-town visitors that day, but unlike the president and his retinue, the press and the bikers, these visitors were not welcome. A Topeka-based group which calls itself the Westboro Baptist Church (WBC) issued a press release days before the anniversary to say that it was planning to picket the president's appearance. The WBC is well-known for its virulently anti-gay views. Its opposition to George W. Bush arises from what the church contends is the way the president is "pretending he's righteous when he's given this country over to the fags," in the words of Shirley Phelps-Roper, daughter of the founder of the WBC, Fred Phelps.[134] Effectively she was a spokeswoman for the WBC.

Phelps maintains a web page entitled "God hates fags," in which the WBC lists its calendar of pickets and press releases. The page relies heavily on fragments of Biblical scripture and it greets its visitors with the salutation, "Welcome, depraved sons and daughters of Adam."

Despite its designation as a Baptist Church, the WBC has no affiliation with any Baptist conference. It is organized around Fred Phelps and members of his family. Trained as a lawyer, Phelps earned accolades from organizations like the NAACP for his work on behalf of civil rights. But his vitriolic tirades about homosexuality and his public protests are what have earned him heaps of media coverage. In 1979, Phelps was debarred by the Kansas Supreme Court which said at the time that the man "had little regard for the ethics of his profession." His daughter, Shirley, is now

---

[134]Mark Anderson, "WBC set to picket Bush, Greensburg May 4," *KCS*, April 30, 2008.

the most publicly visible member of the clan. A lawyer herself like other members of the Phelps family, Shirley is the mother of 11 children. At the tender age of three months, her youngest child participated in one of Phelps-Roper's protests. The babe-in-arms was strapped to Phelps-Roper's back. Proposed WBC picketing, Shirley told Mark Anderson, was directed in equal measure against Bush and the town itself, which she described as "the face of the wrath of God upon this state and on this nation." A banner had already been prepared for May 4, with the picture of a tornado and the words "God's Fury" above it.

The WBC has picketed numerous events in the state over the past ten years. Most often these events had no gay content, but their presence has won the WBC a truckload of publicity. Most recently it has taken to picketing military funerals. It claims to view soldiers' deaths as God's punishment on a nation which tolerates gays. One funeral to be picketed was that of Jessie Davila, a Greensburg native who was killed in Iraq. Davila's funeral was held at Dodge City on March 4, 2006. The WBC showed up to wave upside-down American flags and to hold signs which said "God hates America," "Fag Priest," and "Thank God for 9/11." The Patriot Guard Riders (PGR) were also there to escort Davila's body from Greensburg to the church, and then to the cemetery, acting as a buffer between mourners and the WBC demonstrators.[135]

PGR came into existence in August 2005, growing out of the American Legion Riders Post 136 in Mulvane, Kansas. It arose, in fact, as a direct response to the WBC's picketing at military funerals. Director of the Mulvane post, Chuck "Pappy" Barshney, appointed four colleagues—Terry "Darkhorse" Houck, Cregg "Bronco 6" Hansen, Steve "McDaddy" McDonald, and "Wild Bill" Logan—to come up with a plan to confront, nonviolently, the WBC's stormtroopers. The mission statement of the group states that its objectives are to show respect for those who had given their lives in the Armed Services and to shield the mourning family and their friends from demonstrators like the WBC. Since its creation, the Patriot Guard has spread across the country.

It was sheer coincidence that the WBC and the Patriot Guard were both in Greensburg on May 4, 2008. According to Dodge City biker, Kenny Thomas, the senior class had invited the American Legion Guard to honor

---

[135] The WBC picketed in the vicinity of the funeral of Lance Corporal Matthew Snyder whose father Albert sued for invasion of privacy and the intentional infliction of emotional distress. The case ultimately ended up at the Supreme Court which ruled 8-1 in favor of the WBC. Writing for the majority, Chief Justice John Roberts agreed that speech can inflict great pain but that, under the First Amendment, it must be protected "to ensure that we do not stifle public debate." See Adam Liptak's article, "Justices Rule for Protesters at Military Funerals," in the March 2, 2011 issue of the *New York Times*. The single dissenting justice, Samuel Alito, wrote that "in order to have a society in which public issues can be openly and vigorously debated, it is not necessary to allow the brutalization of innocent victims."

the graduates by holding American flags (right-side-up) along Main Street. Be that as it may, the Guard's presence might easily have been viewed as a deterrent for any WBC protest. The only picketing which occurred involved two adults —one of them Shirley Phelps-Roper, and several kids, presumably her children—who came to wave their signs. The police assigned them a lot three blocks north of the gym where President Bush spoke. There they were guarded (and outnumbered) by the local police. After about ten minutes, they packed their bags and went off. In its press release announcing the group's intention to picket Bush, the WBC described Greensburg as "a God-cursed town . . .[and] rebellious little demon-possessed hotbed of evil masquerading as a municipality."[136] Steve Hewitt dismissed the vituperation out-of-hand: "We don't need negative stuff coming to this town," he said.

> What [Phelps] represents is irresponsible. Greensburg is having an inspiring effect on Kansas and the US with our effort to come back . . . as a green community. . . . After what these kids have been through I just hope the [WBC] presence here doesn't put a damper on the day. I don't think it will. [137]

****

In his commencement address, President Bush had little to say about Greensburg going green. "Greensburg, Kansas, is back and its best days are ahead," he said at one point in his speech. The only mention of the town's commitment to rebuild in an ecologically sustainable way came in his remark that "this community is dedicated to putting the 'green' in Greensburg."[138]

There was a comparison of the resilience of Greensburg residents to rebuild "stronger and better than before" to the country's responses to the terrorist attacks of 9/11 and the devastation of Hurricane Katrina. The president echoed Pastor Blackburn's observation when he said, "The class of 2008 has learned the value of serving a higher cause."

As had happened a year before, the presidential plane, Air Force One, had landed at McConnell Air Base in Wichita. From there, Bush and members of his entourage arrived in Greensburg via the Marine One helicopter. Access to US 54 was blocked from town streets from 1 p.m. until after the president's departure. Unlike his previous visit, President Bush spent no

---
[136] Mark Anderson, "WBC set to picket Bush, Greensburg May 4," *KCS*, April 30, 2008.
[137] *Ibid.*
[138] Mark Anderson, "Bush gives GHS grads a day to remember," *KCS*, May 7, 2008.

time with people outside the high school gym where the graduation ceremony took place. His motorcade sped down Main Street which was lined with Patriot Guards holding American flags. Bush could be seen waving through the bullet-proof windows of his black SUV. Onlookers waved back and cheered although some expressed disappointment that his passage through town was so rapid. The *Signal* quoted one man as saying "Here he comes," followed seconds later by, "and there he goes."[139]

Not surprisingly, extraordinary measures were taken at the gym where the president spoke. To enter was like going through security at an airport. Inside, the press were herded into a couple of rows of chairs at the back of the hall. Behind them, TV and movie cameras were set up on a couple of risers. White House personnel and the Secret Service watched the audience like hawks. Time and time again, people standing in front of exits were shooed away. President Bush was given a standing ovation when he strode on stage. It was shortly after 3 p.m. Camera flashes went off in quick-fire succession around the hall to record the event.

Before Bush spoke, Superintendent Darin Headrick was presented with the "Make A Difference" award for 2007-2008. After his speech, the president presented each of the 18 graduating seniors with a diploma. Bush made his exit before anyone else was permitted to leave. His departure was a repeat-in-reverse of his arrival, this time from the grounds of the high school. A convoy of two white-topped helicopters and three Chinooks spirited Bush back to Air Force One at McConnell Air Base.

As brief as it may have been, Bush's visit was appreciated by the graduates, their friends and family. "I'm very proud Bush is showing some attention to this part of the country," said one woman on Main Street.[140]

****

Any flies in the ointment? In an interview with the *Signal*, John Janssen told Mark Anderson that he had "heard the coffee shop talk of 'I hope Greensburg falls on its face.'"

"The town can still fall on either side of the fence," Steve Hewitt said in September 2008. A month later, he confided that ten percent of the townspeople would follow their municipal leaders while another ten percent—Hewitt called them the CAVE people (citizens against virtually everything)—would oppose the leadership. The 80 percent in between needed to be brought along. These were the residents who would often

---

[139] Kevin McClintock, "Onlookers get only seconds to 'Hail the Chief'," *KCS*, May 7, 2008.
[140] *Ibid*

say "I've never been involved." In the City Manager's view, everyone had to be engaged.

The working relationship between him and Mayor Dixson was another issue of concern. John Janssen offered this opinion on that subject:

> Right now we keep moving the ball down the field, but if somebody fumbles that ball, you could stall right where it is. . . . Bob [Dixson] knows rules and regulations well since he has worked for the post office. And his attitude seems to be 'We'll do the best we can, and if we run out of money, then we'll do something else,' as opposed to 'We're going to do this and do it right and make it work.' . . . . One of the things that concerns me is that he and Steve [Hewitt] don't necessarily work that well together. . . . If they can get . . . in the same harness and pull down the road together, things can happen because Bob's got the time to be mayor, and that's something I didn't have.

City Administrator Steve Hewitt

Hewitt said that all three mayors under whom he had worked were "excellent men." His biggest worry was that people who just wanted to get their lives back together would give in to the pressures of cost and time. The outcome of that struggle, he said, would determine whether he stayed or left.

Seventeen students from Wichita State University came to Greensburg in the spring of 2008 to find and write up stories which had fallen between the cracks of previous reporting on the town. One of their pieces, entitled *A rebuilding rift*, makes the point that Steve Hewitt worried about. In the story, Harry Nolan—the son of Frances Nolan, the woman whom Gary Goodman had helped to rescue on the night of the tornado—says:

> Greensburg is not a hunky-dory place. It's like any other town. You hear all the feel-good stuff, but you don't hear all the not-so-good stuff.

Nolan had come back to Greensburg from Alaska—where he had worked as a software operator—to help out his mother. Another man interviewed for the same article complained:

> We're still trying to get houses back, but we're not getting any businesses back. It's a year down the road and the only thing we've got is the Kwik Shop."[141]

The Kwik Shop was owned by Dillons, a chain of supermarkets. Along with its parent company, Kroger, it had come in for criticism from Representative Dennis McKinney for its failure to immediately rebuild after the tornado. What happened was that Dillons sold the ruin of its old supermarket to the Fraternal Order of Masons. The Masons restored the building while Dillons tacked on a brand new grocery store to the Kwik Shop. The new store was opened in early February 2009, with both Governor Kathleen Sebelius and Secretary of Homeland Security, Janet Napolitano, in attendance. The need to attract business and industry to the town was apparent to everyone. In the September 2008 interview cited above, John Janssen expressed the opinion that:

> We keep getting nibbles, but no one wants to walk through the door. If you think about it, we're in the middle of nowhere, and everybody's hooked on the chicken-and-egg thing. Can I get the employees I need if I build a facility here? My response is that you could easily, but that's my bias.

---

[141] Matt Heilman, "A rebuilding rift," *www.greensburgrebirth.com*, June 6, 2008.

Later in the same interview, Janssen remarked that:

> The purpose of the business incubator is to bring back smaller, Main Street businesses that could not afford rent in a new building. If you're paying $100-$150 a month rent on an 80-year-old building, that's one thing. If you put up new construction and that jumps to $1,500 a month, then you can't support that.

Progress was made on a number of projects that the city of Greensburg and Kiowa County undertook. One of the most striking of these was a rehabilitation of the Kiowa County Courthouse, a building dating back to 1914. Built on a concrete frame with walls two feet thick, brick-skinned but without insulation of any kind, and resplendent with a limestone base and capitals, the courthouse cost $43,500 at the time of its construction. During the tornado, a Pontiac Bonneville had slammed into the southeast corner of the courthouse. Then, sucked up by the wind, it flew over the building, gashing a hole in the roof before coming to rest on the north lawn of the courthouse.

FEMA's *Long-Term Community Recovery Plan* spoke of the need to renovate the building in light of its prominence and cultural value. A decision was made to tear out the walls but to leave unchanged the exterior of the building. A price tag of $5 million was placed on the renovation by Wichita architects MVP, the same group to design, with BNIM assistance, the business incubator. This time the goal was to achieve LEED gold certification. To pay for the renovation, Kiowa County had received close to $2 million in insurance proceeds. Another million came from a variety of sources, including FEMA. Once again, Robin Hood to the rescue, USDA's Chuck Banks made up the funding gap with a contribution of approximately two and one quarter million dollars.

Kiowa County Commissioner Gene West toured the work site in October 2008. The fourth generation of his family to live in south-central Kansas, West took pride in the fact that his great-grandparents homesteaded south of Greensburg around 1884. Sporting a hard hat, he inspected what was being done in the gutted courthouse by Spanish-speaking workers. At the time, insulation had been blown into the roof, and the north facade of the building had been restored. West pointed out a gazebo to the northwest of the courthouse. This doll house of a structure had been splintered by the tornado. National Guardsmen rebuilt it in their free time, making the gazebo one of the first buildings in town to be restored. Adjacent to the gazebo, 32 geothermal wells were being sunk to a depth of three hundred feet into the earth to provide heating and cooling for the new courthouse, which opened for business in mid July 2009.

The townsfolk whom Steve Hewitt had nicknamed the CAVE people were quick to grumble about the cost of the renovated building.

Kiowa County Courthouse under repair

'I can't believe my (tax) money is going to pay for this $5 million thing.' Pepper that statement with a few well-chosen expletives and you get a sense of the kind of comments county officials . . . have been hearing in recent weeks regarding the cost of ongoing and soon-to-be-completed rebuilding projects.[142]

Thus ran the dirge of the CAVE people to going green. In his article, Mark Anderson tried to correct what, in his words, was "the public perception the County is spending taxpayers' money as wildly as the Obama Administration." Anderson quoted a county official as saying that "only $800,000 [of the $5 million project] is coming from the county." The official pointed out that $300,000 was coming in the form of a USDA loan, an additional $315,000 from FEMA and $42,000 from the State's Department of Emergency Management (KDEM). Nearly $2 million came from insurance proceeds, and another $2 million from a USDA Rural Development grant. So there! It was the country's taxpayers—not Kiowa County's—who were making all this possible.

What Kiowa County got for US taxpayers' money was a renovated facility in which all but three of the doors were solid oak portals salvaged from the original courthouse. The floor in the basement was composed of tiles made from crushed glass bottles.[143] Reconstruction was "an unwieldy

---

[142] Mark Anderson, "Funding of projects not likely to burden county's taxpayers," *KCS*, June 26, 2009.

[143] Beer bottles at that. A project manager at MVP said that this was the first time the Wichita firm had used flooring of crushed glass, something called Fritztile. It appealed both as a design element and as an example of recycling.

thing," Gene West admitted. "Rome was not built in a day," he observed philosophically, "but it was destroyed in a day." The County Commissioner said that the July-August utility bill for the courthouse in 2009 came to $1400. Three years earlier for the same period it had run more than twice that at $3400.

Of the new projects, the priciest was the Greensburg school complex. Rising just south of the previous site of the schools, the complex came in at $48 million. That included just about everything on the schools' wish list. Ground-breaking on this project took place on October 29, 2008, with a completion date in mid-2010. The event was attended by dignitaries including Governor Kathleen Sebelius and Representative Dennis McKinney, who told those in attendance that "if we lose our faith, we lose everything." Representing BNIM, the firm which had designed the new complex, architect Casey Cassias said that the schools would "set a new level of education."

Lead architect for the schools design was Joe Keal. At 6 foot 5, Keal holds the distinction of being the tallest person interviewed for this book. A native of southeast Kansas who finds that "the prairie is a beautiful thing," he graduated from KU in 2001 after participating in Dan Rockhill's Studio 804. (The Greensburg story abounds in these connections/coincidences.) Darin Headrick told Joe Keal, who was brought into the project at the beginning of 2008, that the school district wanted "everything for its kids." A new model was developed for the schools, one based on sustainability. BNIM had assistance in designing the school complex from NREL and the USGBC.

Location for the new schools played a role in their design. There were two contending sites, one outside town to the east of Greensburg and another close to where the old high school had stood on Main Street. This latter location was favored by BNIM although it was controversial. The site was close to the flood plain, which had migrated to the northwest since its mapping by FEMA in 1976. The reason for this was that farming practices had altered the nature of the land. BNIM wanted to restore the flood plain to its original state through sensitive landscaping and the use of native plants. In the end, Steve Hardy was able to persuade Darin Headrick to agree to the site adjacent to Main Street.

Each component of the school complex was to have its own distinctive identity. Lower grades—kindergarten through middle school—were housed in the south part of the complex in what looks, from above, like a broken V. The high school, with science labs, was lodged in a long, two-story structure just to the north of the K-7 building in a core one-room deep. The corridor linking the classrooms was designed to be an interaction space. Each classroom was naturally lit, thus eliminating the need for artificial lighting during daytime. Another environmental element—if funding permitted—was water harvesting and a green roof. A common cafeteria and kitchen were located on the ground floor of the high school building. The main gymnasium and indoor track joined the high school on the north side of that building. There were even plans for an outdoor classroom. Concrete masonry units were to be used in the actual construction of the complex, with SIPs for the exterior.

Who was paying for this project? According to Superintendent Darin Headrick, insurance accounted for $16 million. Approximately $21.6 million came from FEMA and $2.8 million from the State of Kansas. This left a gap of $5-10 million, depending on what USD 422 was willing to sacrifice from its construction budget. How was this to be bridged? The rub, as far as FEMA was concerned, was that the Feds would pay 75 percent of costs, after insurance, to put things back as they were. But to return to the *status quo ante* was impractical. As Headrick pointed out, the oldest building in the original complex dated back to 1923. The new complex was to be built to LEED platinum certification. The new schools boasted a slew of green features, even cisterns for irrigation, geothermal heat pumps, and wind generators. Everything gee-whiz and up-to-the-minute. But this was not the first platinum-certified school building in the world—that distinction went to Sidwell Friends School in Washington, D.C., where both the Clintons and the Obamas had sent their kids. In early 2009, Headrick said that the final price tag on the school complex was of the order of $50 million, of which $47.5 million "was in the bank." Shortly afterwards, he gave Mark Anderson the same building cost but said that the school district was negotiating with USDA for the last $4.5 million, in terms of loan money versus grant money. Even with these last figures, USD 422 had secured more than 90 percent of the funds necessary to build.

Headrick offered a couple of examples of the partnerships which Steve Hewitt had hoped to promote in rebuilding Greensburg. The Clorox Company had written a check for $500,000 for the new school complex, and Amazon.com had contributed bikes to 200 children at the end of the 2007-2008 school year. In addition, more than $1 million had come in from 500 different sources.

****

Greensburg officials knew that money could be saved by eliminating redundancies. An example of this involved the new schools' willingness to pass on an auditorium and to use instead the rebuilt Twilight Theatre up the road.

The fate of that building had not been fully settled by the time this book went to press. But hope endureth among the faithful. Farrell Allison, who chaired the Twilight Theatre Board of Directors until mid-2010, expressed confidence that enough money would be raised to reconstruct the building. John Janssen, then treasurer of the Board of Directors, said that they had $600,000 in cash from insurance and tax credits. He reckoned that about $1 million to $1.4 million would have to be found to erect the Art Deco structure, which was designed by the architectural firm of Spangenberg Phillips of Wichita. FEMA could not help in reconstructing what it regarded as a movie house although the Twilight had a non-profit 501(c)(3) status and some stage productions had been done in the Twilight before May 4, 2007. Both Allison and Janssen said that Leonard Burke, who had grown up in Greensburg, had put the Board in contact with Bill Warren, a local movie mogul who was said to own all the movie houses in Wichita and some in Oklahoma. Warren agreed to provide the insides of the new Twilight Theatre—screen, projector and sound system—once the building went up. It was Warren who paid the architect for the project design.

Allison had said—many times over—that construction of the Twilight would not begin until all the funding was in hand. But he told the *Signal* on Feb. 10, 2010, that if the theatre board could raise $70,000 to "complete its share of a $100,000 matching grant from South Central Community Foundation (SCCF)," construction of the building's shell might be undertaken.

But the road to the Twilight, like the yellow brick road in Oz, was a rough one. The last chapter provides a résumé of the pot holes on that road and strategies to avoid them.

Between the sites for the theatre and the school complex, two new buildings were going up on the east side of Main Street. The first of these was the City Hall designed by BNIM and under construction by the late fall of 2008. Its neighbor to the immediate south was a Kiowa County project, the Commons Building. Architects for the Commons Building worked for the Wichita firm MVP, which was involved in other projects in Greensburg cited previously. This county-run facility was ultimately price-tagged at $6.1 million and slated for completion in 2010 although the completion date had to be pushed back to 2011. Securing the funds for the building proved to be a time-consuming process, but construction bids finally went out in November 2009. A date for the ground-breaking was set for one month later. Like the other public buildings under construction in Greensburg, this one was to achieve LEED platinum certification.

The innovative jewel of the Commons Building was a Media Center to be located on its second floor. A press release from *Airspan*, a provider of WiMAX-based broadband wireless access networks, said that it would donate equipment for the Media Center.[144] Emporia-based Stutler Technologies was awarded the contract with Haviland TelCo which would be responsible for the "deployment, operation and management of the countywide WiMAX system."

---

[144] *Airspan* press release, Nov. 19, 2008.

At a technology summit held in the 5.4.7 Arts Center in late June 2008, the *Signal* reported that:

> The wireless 3.65 GHz broadband would reach a 10-mile radius from three base-stations in the county allowing for coverage of virtually 95 percent of the county. It will also provide mobile roaming which will literally make the entire county a "hotspot" for Internet access.[145]

Jan West, president of the board of directors of the Kiowa County Commons' Media Center (and Gene West's wife), spent some time in October 2009 explaining what amazing things the Internet-based TV and radio station at the new center would allow people to do. This included the capability to make personal DVDs. Of the three antennae envisaged for the Media Center, Airspan had already installed one atop the grain elevator in Greensburg. The county would be buying the other two and placing them, one in Pratt and the other at a point on US 183 near the Kiowa/Comanche county line.

Another component of the Media Center was the design and construction, at K State, of a HDTV Remote Broadcast Trailer. The trailer was on display in late October 2010, when students from the Greensburg schools got a tour of the facility. They then got to try out the equipment for themselves. Described as "so sophisticated that anyone can operate it with a little instruction," the cameras were turned over to sixth graders for them to work.[146] Jan West, a retired teacher, was clearly psyched about her new project. "I'm more alive now than I was ten years ago," she admitted.

The most nostalgic feature of the new Commons Building was to be a replica of Dicky Huckreide's famous soda fountain from Hunter Drugs. " We've recreated the original soda fountain where you enter [the building] so people can have a drink if they want to linger for a while," said Tod Ford, an MVP architect who gave a presentation about the Commons Building on October 23, 2008.

The roof for the Commons Building would hold a native-plant garden as well as a system for filtering and saving rainwater. Energy for the system would come from a pair of "egg-beater windmills" in front of the structure.

It would be "completely energy independent and sustainable," Ford told his audience, "able to generate its own power." Ford went on to list some of the green features of the building: photovoltaic panels installed on the roof, light monitors in the rooftop garden, awnings and roof overhangs, and trees to mitigate the effect of light and heat in the summer. The wall construction

---

[145] Janet West, "Integration the key to wiring Greensburg," *KCS*, July 2, 2008.
[146] Jan West, "Media trailer on display," *KCS*, Oct. 27, 2010.

would be similar to what had been used on the Business Incubator on the other side of Main Street.

As in the case of other Greensburg building projects, funding for the new building came from a variety of sources: insurance reimbursement, FEMA, the USDA, and corporations to be button-holed by then-Governor, soon-to-be Secretary of Health and Human Services, Kathleen Sebelius. "Every little bit helps," observed Ford.[147]

Positive response to Tod Ford's presentation, however, had not been universal. In an interview with the *Signal*, ex-mayor John Janssen complained:

> Couldn't the library add another layer of utilization and cooperation at the Big Well? Would it be appropriate to locate the K State extension office in this Big Well facility? . . . Once the FEMA, State, USDA grants and insurance cash is gone, the taxpayers of Kiowa County will be paying for these facilities. In the words of the philosopher, 'Less can be more.'[148]

When asked if he thought that the advantages of placing the museum, the library, and the KSU extension office in the Big Well complex outweighed the programming possibilities of linking them with the Media Center, Janssen became even more expansive:

> I know the Internet was available in the library, though I don't think they had much to do with it in the museum. But with all three entities, the key to success . . .is the human touch. If I go to the library, extension office or museum and have a good experience with the people there, that is word-of-mouth promotion. If the Internet is the key, people will stay home.[149]

Anderson asked Janssen what he missed most about being mayor, and the tax consultant-cum-rancher responded:

> What I miss most is being in the game. I love a puzzle and Greensburg was a good one. I'm not a very good politician, so I don't miss trying to be one. My wife is thrilled to death I'm not mayor because I'm in the office now generating cash flow. She likes that.

---

[147] Mark Anderson, "Commons design a hit," *KCS*, Oct. 29, 2008.
[148] Mark Anderson, "Though out of office, Janssen still has plenty to say," *KCS*, Aug. 1, 2008.
[149] *Ibid.*

REALITY CHECK

"Tell me something about yourself, and the country you came from," said the Scarecrow. . . . So she told him all about Kansas, and how gray everything was there.

–Wonderful Wizard of Oz

On April 19, 2009, two journalists visited Greensburg. There was nothing so unusual about that. Media types were always coming to town. It wasn't even so unusual that they were not American. Greensburg was getting more and more attention from European and Japanese publications. What made the trip by Steffen Kretz and his cameraman unique was that these two men from the Danish Broadcasting Corporation had included Greensburg on their whirlwind tour of the States as part of a program to be aired the month preceding the Copenhagen conference on climate change. That took place in December 2009. This UN-sponsored event—called the COP15 Conference—had for its objective the recasting of emission regulations which were introduced first at a conference in Rio de Janeiro, 1992.[150] The Kyoto Protocol, adopted in Kyoto, Japan, on December 11, 1997, set binding targets to reduce greenhouse gas emissions for the earth's industrialized countries. Kretz, a well-known journalist in his native Denmark, wanted to see what was happening in the US with regard to environmental sustainability. His half day in Greensburg began on a disappointing note. He had expected the whole town to look like the 5.4.7 Arts Center which clearly grabbed his attention, almost to the exclusion of everything else. But as the hours ticked past, Kretz came to recognize and admire what the town had managed to accomplish.

---

[150]The name COP15 stands for the fifteenth meeting for the Conference of the Parties, the supreme body of the UN Framework Convention on Climate Change. The Copenhagen Conference was judged to be a failure. See Bill McKibben's assessment in the chapter *The Context*.

For much of the world, Greensburg was the green beacon on the hill. But paradoxically the *Kiowa County Signal* ran a series of articles which savaged the foundation of the town's sustainable development. These began to appear in late 2009. They suggested that anthropogenic contributions to climate change—global warming, in particular—were a hoax. "Global warming hoax uncovered," screamed the headline of the first of these articles on December 2, 2009. Along with the articles came some vicious cartoons, one impugning the integrity of climatologists, another with Nazi-looking cops sporting EPA armbands and berating a man labelled "Public," and yet another portraying Al Gore as a bloated-paunch liar.

Source of these op-ed pieces was an organization called *Americans for Limited Government* (ALG) and its affiliated news service, *Liberty Features Syndicate*. ALG is a non-profit 501c(3) organization which, according to the nonpartisan and nonprofit Center for Public Integrity, received 99 percent of its funding (in 2005) from three unnamed sources.[151] The Center claimed that these donors were connected to ALG's chairman, New York real estate investor Howard Rich.

Prominent among climate skeptics in this country are two brothers, Charles and David Koch, whose family-owned company—Koch Industries—is headquartered in Wichita, 100 miles east of Greensburg. The brothers, who are among the ten wealthiest men in America, have provided funding for a number of right-leaning organizations. One of these, the Cato Institute, a Washington, D.C.-based think tank, received $11 million from the Koch brothers over a period of seven years, again according to the Center for Public Integrity. Cato has been relentless in its criticism of efforts to counter global warming. The brothers' support for organizations like the Cato Institute is widely viewed as self-serving since Koch Industries operates oil refineries in three states and controls four thousand miles of pipeline.[152] The tactics which the brothers employ in pursuit of their political agenda has prompted outright condemnation in some circles. Charles Lewis, the founder of the Center for Public Integrity, said of them:

---

[151] www.publicintegrity.org/investigations/takings/articles/entry/69

[152] The brothers try to avoid the public spotlight, but with less and less success as their notoriety grows. They and their company have recently been the subject of a spate of articles, one of which is cited in this paragraph. Another article appeared in a Greenpeace publication available online at *greenpeace.org/kochindiustries*. Koch Industries retorted that the Greenpeace report 'distorts the environmental record of our companies.' *Rolling Stone* cited the brothers in two consecutive years, first in an article entitled "The Climate Killers" which appeared on January 21, 2010, and again on February 3, 2011, in an article by Jeff Godell entitled "Who's to Blame." In that article, the brothers ranked Number 2, between Rupert Murdoch and Sarah Palin, in a list of the twelve "politicians and corporate executives most responsible for blocking efforts to halt global warming." A meeting of conservatives, hosted by the Kochs in Southern California the weekend of January 30-31, 2011, was the occasion of a counter-demonstration numbering close to 1000 activists. Greenpeace hired a blimp emblazoned with images of the Koch brothers and the phrase "Dirty Money."

They have a pattern of lawbreaking, political manipulation, and obfuscation. I've been in Washington since Watergate, and I've never seen anything like it. They are the Standard Oil of our times.[153]

The incident which launched the series of articles in the *Signal* came to be known as Climategate. It began when parties unknown hacked the server of the Climate Research Unit (CRU) at the University of East Anglia in late November 2009. The hacker(s) uploaded 1,000 e-mails and numerous documents to a Russian computer site before distributing them widely across the Internet. Skeptics pounced on words like "trick" and phrases like "hide the decline" to claim that a clique of climatologists was perpetrating a hoax on the public. Senator Inhofe, whose own views tallied with those expressed in the *Signal* articles, lost no time in crowing that "ninety-five percent of the nails were in the coffin prior to this week; now they are all in."[154] What made the leaked e-mails significant, according to ALG, was that the CRU data sets were "widely used in climate research, including the global temperature record used to monitor the state of the climate system." IPCC's data for its fourth assessment report had been compiled, it was claimed, by the CRU, and its projections for temperature increases looked increasingly dubious.[155]

Cato spokesmen had a field day with Climategate. One of the institute's scholars "gave more than twenty media interviews trumpeting the alleged scandal."[156] The leaked material from the CRU did cast an embarrassing light on the all-too-human nature of some of the authors of the purloined e-mails, but for all the brouhaha that it generated, Climategate did not establish that global warming was left-wing scare-mongering. The University of East Anglia issued a statement which said that "the selective publication of some stolen e-mails and other papers out of context is mischievous."[157] Head of the CRU, Phil Jones, acknowledged that "some of the published e-mails do not read well," but he cautioned that the flap should not overshadow the "enormous challenges ahead." He went on to comment:

---

[153] Jane Mayer, "Covert Operations," *The New Yorker*, August 30, 2010.
[154] Kimberley A. Strassel, "Cap and Trade is Dead," *Wall Street Journal*, Nov. 26, 2009.
[155] On Nov. 15, 2010, the *Washington Post* ran a Reuters release stating that much of the COP15 was taken up by wrangling over the hacked e-mails from the CRU. If the aim of the perpetrators of the leak had been to throw a monkey wrench into the conference proceedings, they succeeded. The same Reuters release asserted that "less than ten percent of the news articles written about [the conference] dealt primarily with the science of climate change."
[156] Jane Mayer, "Covert Operations," *The New Yorker*, August 30, 2010.
[157] David Stringer, "Hackers leak e-mails, stoke climate debate," *Associated Press*, Nov. 21, 2009.

That the world is warming is based on a range of sources: not only temperature records but other indicators such as sea level rise, glacier retreat and less Arctic sea ice.

Our global temperature series tallies with those of other, completely independent, groups of scientists working for NASA and the National Climate Data Center in the United States, among others. Even if you were to ignore our findings, theirs show the same results. The facts speak for themselves; there is no need for anyone to manipulate them.[158]

The *Signal* failed to run any expression of consensus climatology, nor any opinion such as Jones'. Nor did the *Signal* report on subsequent events in 'Climategate,' such as the exoneration of the CRU and its head, Phil Jones, by the British House of Commons. This was made public on March 30, 2010. In its summary, the Science and Technology Committee of the Commons, wrote that "We are content that the phrases such as 'trick' or 'hiding the decline' were colloquial terms used in private e-mails and the balance of evidence is that they were not part of a systematic attempt to mislead."

For its part, the IPCC requested a group called the InterAcademy Council to conduct a review of its procedures. The Council, which consists of presidents of 15 academies of science and comparable organizations from the US, the UK, Australia and other countries released its own findings on August 30, 2010. At a press conference the same day, IPCC Chair Rajendra Pachauri told reporters that "Science thrives on honest, well-reasoned debate." He then went on to caution that "honest scientific discourse wilts under gross distortions and ideologically-driven posturing."[159]

The Council review was the seventh that year. None of the previous reviews had found flaws in the fundamental science of climate change. Pachauri cited the findings of a number of these reviews in support of IPCC work. But climatologists were still not out of the woods. There was speculation that some of them might be subpoenaed to appear before a House of Representatives Committee to be chaired by Darrell Issa of California, after the Republican Congressional victories of 2010. As reported in the November 22, 2010, issue of *The New Yorker*:

At the top of [Issa's] list are the long-suffering researchers whose e-mails were hacked last year from the computer system of Britain's University of East Anglia. Though their work has been the subject of three separate 'Climategate' inquiries— all of which found that allegations of data manipulation were

---

[158] Phil Jones, CRU Update, Dec. 1, 2009, available at www.uea.ac.uk
[159] www.un.org/News/briefings/docs/2010/100830_IPCC.doc.htm

unfounded—Issa isn't satisfied. 'We're going to want to have a do-over,' he said recently.[160]

The InterAcademy Council review did pick up on one of the errors in AR4, to the effect that there was a high probability that the Himalayan Glacier would disappear by 2035 if the planet continued to heat up at its 2007 rate. IPCC was slow to correct this mistake, but one error in a technical document of approximately 3000 pages is in the same league as boo-boos in the Encyclopedia Britannica.[161] "Highlighting this error to undermine climate science, however, is a classic example of cherry picking—a dangerous game to play with 500 million livelihoods at stake."[162]

Pachauri, elected chair of the IPCC in 2002, succeeded Robert Watson, who criticized AR4 for mistakes "making it seem like climate change is more serious by overstating the impact."[163] His remarks surprised some for two reasons: First, because it was believed that a memo by Exxon-Mobil to the Bush administration led to lobbying by that administration to oust Watson and to replace him by Pachauri, who was viewed as more industry-friendly than Watson.[164] Second, because some critics have faulted the IPCC for underestimating the dangers of climate change. A study which appeared at about the same time as AR4 pointed out that temperatures and sea levels have been rising faster than predicted in the Panel's 2001 report.[165]

Probably the highlight of Pachauri's chairmanship was the awarding to the Panel of the 2007 Nobel Peace Prize which the IPCC shared with Al Gore, an association that was unlikely to enhance the prestige of the UN group among the residents of Greensburg. But to return to Point A, the decision by the *Kiowa County Signal* to give such prominent play to the climate deniers and to ignore—entirely—subsequent investigations of various aspects of Climategate raises an obvious question: How—in a community committed to sustainable development—could this happen?

---

[160] Elizabeth Kolbert, "Talk of the Town," *The New Yorker*, Nov. 22, 2010.
[161] www.skepticalscience.com/IPCC-Himalayan-glacier-2035-prediction.htm
[162] *Ibid.*
[163] Ben Webster and Robin Pagnamenta, "UN must investigate warming 'bias', says former climate chief," *The Times of London*, Feb. 15, 2010.
[164] See Fred Pearce, "Top climate scientist ousted," *New Scientist*, April 19, 2002, and Julian Borger, "US and Oil Lobby Oust Climate Change Scientist," *The Guardian*, April 20, 2002. Pearce wrote that Watson "fell victim to an alliance between the US—which may have threatened to drop funding for the IPCC if he stayed on—and many developing countries, who took the chance to appoint 'one of their own' to the world's top climate science job." Pearce went on to quote Kate Hampton, a climate coordinator for Friends of the Earth, to say that "the Bush administration and its friends would rather shoot the messenger than listen to the message." Borger wrote that "environmentalists argued that the anti-Watson campaign was a show of strength by the US, oil producers like Saudi Arabia, and oil corporations like Exxon-Mobil, intended to cow the IPCC."
[165] See Richard Blake, "Humans blamed for climate change," Feb. 2, 2007, BBC News, and Bill McKibben, "Warning on Warming," *New York Review of Books*, March 15, 2007.

****

Professor Chris Crandall's fourth-floor office at the University of Kansas looks west over Jayhawk Boulevard. It is furnished with chairs designed by Charles and Ray Eames. His bookcase, another Eames creation, contains books about their furniture. The ceiling is covered in rolls of cloth baffling which hide the nicotine stains, a legacy of the previous occupant. The billowing material also softens the light of the fluorescent neon tubes affixed to the ceiling. A bicycle which Crandall uses to commute between his office and home five miles away lies propped against the wall.

"Your problem," Crandall said, "is to find the root belief" of people like Mark Anderson, editor of the *Signal*. "People feel the need to return to their core beliefs which are interconnected."

Conversation turned to what had already been accomplished in Greensburg—the buildings which had been completed, were under construction, or out to bid, all of them meeting a high LEED certification. To trash-talk a principal concern of the green movement at this point was like shutting the barn door after the cow had got out, even if criticism of the sort appearing in the *Signal* could impact private construction in town. "The time to make noise is when the money is in the pipeline," Crandall observed wryly. He went on:

> Why are people making noise which is contrary to their self-interest? There may come a time when self-interest actually disappears in [a context of] politics and religion.

Townsfolk react to a stimulus, *e.g.* going green in Kiowa County and the cost of going green, in terms of their pre-existing views and values. Greensburg was collectively on the political right before the tornado. You would expect it to stay on the right in the aftermath of the tornado. All that the Department of Agriculture had done for Greensburg was good, yes, but big government spending was bad. There was the rub. It might be instructive, said Crandall, whose expertise is social psychology, to view Mark Anderson's fit of establishment-climatology bashing in terms of *cognitive dissonance*—a feeling of tension which results from retaining in the mind two conflicting thoughts or ideas at the same time.

As for Al Gore, a favorite target of climate-change deniers: "He changed the debate on climate almost single-handedly," in Crandall's opinion. Perhaps because of this, Gore couldn't get elected dog-catcher in Greensburg. His arguments about anthropogenic climate change are commonly ridiculed on the prairie. But the Bible instructs us to go green, to be good stewards of God's earth. And doesn't it make sense to economize and preserve our natural resources? *Voilà* a resolution of the dissonances, cognitive or otherwise. Some reasons to strive for sustainability are rejected, but others are found, and—may one say this?—the result is the same.

****

The paved road south of Salina gives out. You drive for a mile and a half, on dirt and gravel, before coming to it: a low-lying cream-colored brick bungalow. Nothing out of the ordinary, a fairly typical Kansas country house, in fact. On the other side of the road is a parking lot of sorts. A half dozen vehicles are parked there. Below the house stands what looks like a barn—the Big Barn, in fact—and, largely hidden by the house itself, a substantial greenhouse.

Inside the house, however, the resemblance to a farm dwelling comes to an abrupt end. An etching of Aldo Leopold, author of *The Sand County Almanac*, a famous environmental tract, holds place of honor in the entry hall. Offices and meeting rooms extend in all directions, instead of the living/dining/bedrooms you might expect to see. In fact, what goes on here is well out of the ordinary. Welcome to The Land Institute—known simply as "The Land" to its many admirers—the headquarters and research labs of Wes Jackson. Like Lester Brown, a MacArthur Fellow, the recipient of numerous awards and honorary doctorates, listed by *Smithsonian* in the fall of 2005 among the 35 innovators "who made a difference," Wes Jackson is one of the most influential figures of the late twentieth/early twenty-first centuries.

Who is he? Cofounder of The Land in 1976, Jackson is a plant geneticist with an all-consuming mission in life. Prone to expressing himself in Biblical terms, Jackson claims that farming methods dating back 10,000 years are humanity's original sin. Ten millenia ago, in an allegorical parallel to the expulsion of Adam and Eve from the Garden of Eden, there occurred:

> The first moment when we began to erode the ecological capital of the soil. That's when humans first started withdrawing the earth's non-renewable resources. It wasn't intentional. It didn't require a chamber of commerce or the devil to make us do it—we just did it.[166]

Ken Warren—a man with a background in geology and a colleague of Jackson's for the past 14 years—said that the goal of the Land Institute was not a small fix, "not little tweaks here and there," but the actual overthrow of agriculture as we know it. Difficulties are long-standing: the

---

[166] Craig Canine, "Wes Jackson," *Smithsonian*, Vol. 36, No. 8, Nov. 2005.

Biblical fall was the beginning of our problems in agriculture. "Our attitude has [always] been 'What can you force out of a place' instead of 'What will Nature require you to do?'" he said. "Nature must become a partner rather than a force to be competed with."

For the past 34 years Jackson and his staff—The Land now employs thirty people, including seven Ph.D.s—have worked to hybridize annual crops like wheat and sorghum with related perennial plants. The idea is to produce crops as resistant to extreme weather as the plants that grow naturally on the prairie. "We have sort of a crash program to develop these crops—if you can have a crash program that's going to take decades," The Land's director of research told the *Smithsonian* in 2005. "The timeline we're working on shows us having a set of perennial grain-producing crops that would be usable in agriculture somewhere between 25 and 50 years from now."

Besides recycling soil nutrients, the perennials' sturdy roots would keep the topsoil from running off. There is an amazing photograph which visitors to The Land all see: It shows the relatively puny root system of domestic wheat, side by side with the tangle of roots—like the tentacles of a super-sized Portuguese man-o'-war—of perennial wheatgrass. These roots reach four times as deep into the soil as those of their kith and kin, domestic wheat.

As Ken Warren explained it, the solution to the problem of feeding the world's expanding billions is to "perennialize" the major grain crops and to grow them in a prairie-like mixture. Eleven of thirteen major grain crops, like wheat and rice, have perennial relatives. The trick is to increase seed production among the perennials through the hybridization techniques being developed at The Land Institute. Advantages abound. With perennials, there would be no need for herbicides or pesticides. You could also cut back on fertilizers, a big consumer of petroleum and a pollutant in aquifers. Warren said that the Kansas towns of Abilene and Garden City were currently spending millions of dollars to extract fertilizer-based nitrates from their municipal water supplies.

It was only natural to wonder what Greensburg residents would make of The Land. How to get a handle on that question? Rex Butler—who maintained an organic garden—was a natural to lure to Wes Jackson's research facility. Arranging this trip was tricky: Ed Schoenberger had retired, and Butler was the freshly minted sexton at the cemetery. If someone had died a day or two before the trip, the mission would have had to be scrubbed to let the new sexton dig a grave. In fact, this nearly happened, but the window of opportunity stayed open long enough to make the three-hour drive north to Salina before Butler had to pull out the back hoe.

It quickly became apparent that Butler was in agreement with what he heard from Ken Warren. Shown a map of how, with global warming, wheat production by 2050 was projected to shift northward deep into the Canadian plains, Butler stated that "we don't know what sustainable means." Warren was clearly relieved to be conversing with someone who understood the problem. This was not always the case. "Why don't we hear more people talking like you?" he asked wistfully. Later in the session, Warren said that "we're not out to please people, just feed them." For his part, Butler confessed to having learned something at The Land. "If perennials are the way to go, then The Land Institute is way ahead of the game," he said. With the price of diesel fuel and fertilizer rising, going perennial made sense. No, it was not a waste of money to support The Land's work—this from a man noted in Greensburg for his frugality.

The following day, back in Greensburg, a casual encounter with a farmer revealed the depth of hostility to the idea of hybridizing sorghum and Johnson grass. The farmer would have none of that. Much of his life in the fields had been spent eradicating Johnson grass. Clearly Rex Butler's perception of sustainable agriculture was not shared by all his neighbors.

****

Besides its research projects, The Land supports a well-attended teaching program. Jackson himself now spends the lion's share of his time lecturing at universities all over the world. Each summer The Land sponsors what it calls its Prairie Festival. Devotees come, year after year, to overflow the Big Barn where speakers from all walks of life preach sustainability. In 2007, then-Lieutenant Governor Mark Parkinson argued—to the delight of his audience—against coal-fired power plants. This was a white-hot issue at the time.

Parkinson went on to decry the lack of political will which had resulted in an increase of greenhouse emissions which led, in turn, to climate change. With rising demand from an expanding middle class worldwide, government was reluctant to raise the cost of energy with schemes like cap-and-trade. His solution? Science. "I really believe that," he told his audience. Funding for research in alternative energy should be increased. Whatever new technology emerged should be shared with the world. He concluded with a favorite quote from Jack Kerouac's *On the Road*:

> The only people for me are the mad ones, the ones who are mad to live, mad to talk, mad to be saved, desirous of everything at the same time, the ones who never yawn or say a commonplace thing, but burn, burn, burn, like fabulous yellow roman candles exploding like spiders across the stars and in the middle you see the blue center light pop and everybody goes 'Awww!'[167]

---

[167] Jack Kerouac, *On the Road*, first published by Viking Press, 1957, now available as a Penguin Classic on Amazon.

Seven hundred people pre-registered for the Prairie Festival of 2010, and more were still streaming in at midday on Friday, September 24, 2010, when the Festival got underway. The crowd—this was the biggest audience ever for the Festival—parked their vehicles in a field beside the lot reserved for The Land Institute personnel. To judge from the license plates, most of the attendees came from Kansas, but some had driven from all the neighboring states for this once-a-year event. They came in all conditions and states of life: elderly environmentalists in wheelchairs and walkers; youths in jeans and T-shirts. One woman with a mop of reddish brown hair swept past, dressed in beaded sandals, yellow PJ's with red ankle bands and a purple-pink tank top. A student in the crowd talked about his interest in the biological aspects of anthropology. Later he confessed that his favorite TV shows were Dexter and CSI. One Mennonite couple sat in the full glare of the autumn sunlight. Had they come from Greensburg?

People were eager to hear the kick-off speaker, Wendell Berry, a poet, novelist, and essayist whose reputation as a spokesman for the environment is of mythic proportions. Born in Kentucky in 1934, Berry is closely identified with the state where he currently lives and farms. The desecration of the state's land by King Coal and the industrialization of agriculture which occurred after World War II was the focus of his talk.

The line-up of afternoon speakers kept the attention of the audience clustered in and around the Big Barn. One speaker whose talk was directly relevant to the concerns of this book was Josh Farley, an ecological economist from Vermont. Farley spoke of the need to radically redesign our economy. The conventional view was that there was no limit to growth, but every economic activity has its cost. We need to transform our goals to protect future generations. People are not insatiable, he argued. For ninety percent of human history, accumulation meant death. Market forces? Farley paraphrased John Maynard Keynes to say that market economics is the astonishing belief that the wickedest men will do the wickedest things for the greatest good of everyone. The free market economy glorifies greed and selfish behavior. We need to reward cooperation. We are leaving the era of fossil fuels. What is needed right now is a kind of Bretton Woods agreement for all essential non-market resources.

The purpose of this pilgrimage out to The Land Institute was to interview Wes Jackson. Did he have an opinion of what the people in Greensburg were doing?

Jackson—a husky man with a resonant baritone voice and a legendary amount of energy—began his answer by noting that 99 percent of all the oil to be consumed by humanity has been burned in his lifetime. (In the spring of 2009, he was 73.) As a people, we have no idea how fossil fuels dictate the patterns of our lives. Greensburg has been presented with the opportunity to make resilience its goal, but Jackson added a serious caveat: You can have your "groovy architecture," and go after "your brownie points"—a reference presumably to LEED certification—but if you don't address the fundamental question of economic growth—if, for example, you open businesses in the expectation that they will grow—then you have "missed the point." That being? "Growth is the problem. We must be the first species to say 'no.'" The belief that growth is necessary has become a kind of religion. If people's thinking involves growth, then "they've missed the point."

Jackson compares the planet's resources to a library with a fixed amount of space. Every shelf is already bulging with books. To add a book, you will need to throw away a book. That is the library of a zero-sum game.

What about the naysayers, like British-born Freeman Dyson? Jackson said that he could admire the intellectual playfulness of the physicist, but he thought that Dyson was still a nut case whose "playful mind [had] run amok."

****

On April 2, 2007, the Supreme Court decided an issue soon to impact the State of Kansas. By a 5-to-4 vote—a tenuous liberal majority against the usual, less-tenuous conservative minority—the Court found that the EPA had "the statutory authority to regulate emission of [greenhouse] gases from new motor vehicles." In deciding the case, Massachusetts *et al. v.* EPA *et al.*, the Court's majority noted that "the harms associated with climate change are serious and well recognized." Furthermore, "given EPA's failure to dispute the existence of a causal connection between man-made greenhouse gas emissions and global warming, its refusal to regulate such emissions, at a minimum, 'contributes' to Massachusetts' injuries." The syllabus (headnote) of the Court finding went on to assert that: "While regulating motor-vehicle emissions may not by itself *reverse* global warming, it does not follow that the Court lacks jurisdiction to decide whether EPA has a duty to take steps to *slow* or *reduce* it."

Writing for the majority, Justice John Paul Stevens began his opinion by noting:

> A well-documented rise in global temperature has coincided with a significant increase in the concentration of carbon dioxide in the atmosphere. Respected scientists believe the two trends are related. For when carbon dioxide is released into the atmosphere, it acts like the ceiling of a greenhouse, trapping solar

energy and retarding the escape of reflected heat. It is therefore a species—the most important species—of a 'greenhouse gas.'[168]

The plaintiffs in the case had charged the EPA with abdicating "its responsibilities under the Clear Air Act to regulate the emissions of four greenhouse gases, including carbon dioxide." Justice Stevens wound up his decision thus:

> EPA said that a number of voluntary executive branch programs already provide an effective response to the threat of global warming, . . . that regulating greenhouse gases might impair the President's ability to negotiate with 'key developing nations' to reduce emissions, and that curtailing motor-vehicle emissions would reflect 'an inefficient, piecemeal approach to address the climate change issue.'
>
> Although we have neither the expertise nor the authority to evaluate these policy judgments, it is evident they have nothing to do with whether greenhouse gas emissions contribute to climate change. Still less do they amount to a reasoned justification for declining to form a scientific judgment.[169]

Within the year, this Supreme Court decision influenced an energy-generation question posed in the State of Kansas. Secretary of the Kansas Department of Health and Environment, Roderick L. Bremby—the same man who would later participate in the ground-breaking ceremonies for the hospital in Greensburg—denied an air permit for a proposed coal-fired power plant. On October 18, 2007, Secretary Bremby wrote: "I believe it would be irresponsible to ignore emerging information about the contribution of carbon dioxide and other greenhouse gases to climate change and the potential harm to our environment and health if we do nothing."[170]

His decision was one of several in the nation which built on the Supreme Court decision the preceding April. In the case of Kansas, the issue dealt with the proposed construction of two 700-megawatt, coal-fired power plants which Sunflower Electric Power Corp. proposed to build in a town 96 miles west of Greensburg. The company intended to build the plants to sell electricity to the neighboring states of Colorado and Texas. According to *Earthjustice*, an advocacy group representing the Sierra Club (which

---

[168] Justice J. Stevens, "Massachusetts *et al.*, Petitioners *v.* Environmental Protection Agency *et al.*, No. 05-1120, April 2, 2007.
[169] *Ibid.*
[170] David Biello, "Coal-Friendly Climate Changes in Kansas," *Scientific American*, Oct. 19, 2007.

opposed the project), the plants would have emitted in excess of 10 million tons of carbon dioxide every year. Said an attorney for *Earthjustice*: "Today, Kansas embraced a bright, clean energy future powered by new technologies that will breathe life into our economy, and took a giant first step toward protecting our children and grandchildren from the devastating impacts of global warming."[171]

Sunflower wasted no time in trying to overturn Bremby's decision. Early in 2008, a bill which would have allowed the coal-burning plants to go ahead and—at the same time—would have restricted Bremby's authority to block permits and to rewrite emissions standards without the Legislature's approval was sent to Governor Sebelius. She promptly vetoed it. Sebelius did say that she would not object to the building of one power plant if Sunflower would make a commitment to capture greenhouse emissions and to develop wind power. Sunflower said that this was financially infeasible.

Environmentalists were ecstatic at Sebelius' stand. A spokeswoman for the Sierra Club said that this action "sends a message that Kansas is willing to be part of the solution, rather than part of the problem." She added, "Kansas is sort of ground zero right now for the global warming debate."[172]

Dennis McKinney—a Democrat, recall—had voted for the bill which Sebelius vetoed. When asked about this, McKinney said that the use of coal was a stopgap measure, something that was necessary until alternative energy sources could be brought onstream. He also made the point that Secretary Bremby could not impose stricter emission standards than federal guidelines allowed.[173]

Sunflower estimated that the plants would produce 11 million tons of carbon dioxide annually, in line with the quantity projected by the Sierra Club. As a sop to environmentalist sentiment in the state, the power company did propose a bioenergy center to capture the majority of these emissions, but it refused to run up the white flag on the issue of power plant construction. Over the next 12 months, three more bills in support of the

---

[171] Nick Persampieri, "Kansas Rejects Massive Sunflower Coal-Fired Power Plant," *www.earthjustice.org*, Oct. 18, 2007.

[172] "Kan. Gov. Vetoes Coal-Fired Power Plants," *www.cbsnews.com*, March 22, 2008.

[173] In 2002, the State of California adopted a law which required reductions in greenhouse gas emissions from light-duty vehicles. These restrictions were more stringent than federal guidelines, and California needed to acquire a waiver from the EPA before the law could take effect. Once the waiver was granted, the state and a number of other states in the Far West, the Northeast and Florida, which had adopted or had announced their intention to adopt the California standards, could proceed to enact them. In December 2007, Governor Arnold Schwarzenegger was informed by the EPA that the state's request for a waiver was being denied, even though EPA lawyers and scientists had all but unanimously recommended that the waiver be granted. Upon assuming office, President Barak Obama directed federal regulators to revisit the issue and to move swiftly on the California waiver request. On June 30, 2009, the EPA did grant the waiver.

power plants passed the Legislature and went on to the governor, who vetoed all of them. On the occasion of her fourth veto, Sebelius said "What was a bad idea last year, is an even worse idea today." Supporters of these measures had the votes to override Sebelius' veto in the Senate but not in the House.

In the spring of 2009, Sebelius joined Barack Obama's cabinet as his Secretary of Health and Human Services. Lieutenant Governor Mark Parkinson—the same Mark Parkinson who had spoken against coal-fired power plants at the 2007 Prairie Festival—became Governor of Kansas.

Within hours of becoming governor, Parkinson undertook negotiations with Sunflower Electric Power to craft a bill that called for the construction of a single 895-megawatt power plant with provisions to create two transmission lines, each capable of carrying 1,000 megawatts of electricity to Colorado. Sunflower agreed to take steps to offset the new plant's carbon dioxide emissions; to build wind-powered generators; and to use biomass fuels as well as coal (although this part of the agreement becomes null and void if the use of biomass is either technologically or economically infeasible). The bill passed the legislature with only token opposition.

Parkinson claimed that the settlement was "a win" for Kansas. Kansas Democratic Party chair, Larry Gates, lauded the governor's leadership in securing a deal which provided the state with a "smaller, cleaner and more efficient power plant" at the same time it provided "new wind farms, transmission lines, net metering and renewable portfolio standards."[174]

Transmission lines were, in fact, essential to the distribution of any wind-generated electricity. Late in 2009, Kansas was shocked to learn that a project to build a major line might be scuttled. The project was restored, but only after frantic lobbying by Gov. Mark Parkinson. Writing in the *Kansas City Star*, Steve Everly observed that:[175]

> The need to improve the [national power] grid is pressing enough that it even has some new champions—environmentalists who once opposed high-voltage lines because they might carry electricity from coal-fired plants. "If we don't have the infrastructure, we don't get the wind industry," said Maril Hazlett, associate director of the Climate & Energy Project, part of The Land Institute.[176]

---

[174]Larry Gates, KDP Chair, "A standstill has ended," an e-mail message sent out May 5, 2009.
[175]Steve Everly, "Lack of power lines a blow to wind energy," *Kansas City Star*, Nov. 11, 2009.
[176]*Ibid.* In light of Greensburg's long and largely futile attempt to attract industry to the new town and its own development of wind-generated power, the construction of a high-voltage transmission line assumed increasing importance as time wore on.

Environmental groups were unenthusiastic about Parkinson's compromise with Sunflower. Lawrence-based *Sustainability Action Network* put the best possible face on developments by reporting on May 5, 2009, that:

> It's always disheartening when political expediency determines policy rather than ethics, science, or conviction. But that said, the deal that Governor Parkinson made with Sunflower Electric has quite a few up-sides to counteract the construction of the one coal plant he agreed to.[177]

The Kansas chapter of the Sierra Club was less forgiving: "While the country is moving away from polluting fossil fuels, Kansas has opened the door for outdated, dirty technology other states are rejecting."[178]

To put the matter in perspective, it should be recalled that Parkinson's deal with Sunflower bore a strong resemblance to what Kathleen Sebelius had said in March 2008, that she would accept (and which Sunflower had declared to be financially infeasible). Sunflower delayed its application for an air quality permit until early 2010. The Sierra Club speculated about this: Did the delay reflect financing difficulties? The prospect of oversight by the utility that would actually own the plant? Or the hope that a decision could be postponed until a new state administration took office?[179]

Called the Kansas V-Plan, the project involved two segments, the first from the town of Spearville to a substation in Clark County, the second eastward across Kiowa County toward Wichita. The two segments formed a 'V', hence the name of the plan. According to Dennis McKinney, an alternative route was scrapped under pressure from the Sierra Club, anxious to protect the habitat of prairie chickens. There were significant financial advantages for the counties through which the line passed: Money paid proprietors to secure the easement for the transmission line; tax revenue coming to the counties through which the line passed; and finally the enticement to build wind farms to feed into the 345 kilovolt line. A wind farm in Greensburg would arise at the southern end of a string of such farms which ran northward to Spearville and beyond. The coal-fired power plant at Holcomb was projected to feed into an extension of the V-Plan from Spearville toward Denver, Colorado.

[177] Michael Almon, "Kansas Coal Plant Compromise," *Sustainability Action Network*, May 5, 2009.

[178] Scott Rothschild, "Environmental group blasts coal plant plan," *Lawrence Journal-World*, May 6, 2009. An intriguing article entitled "Dirty Coal, Clean Future," appeared in the December, 2010, issue of *The Atlantic*. Its author, James Fallows, is a former speech writer for Jimmy Carter. Fallows argues that the world's dependence on coal is certain to persist "for a very long time." The environmental challenge, given such a premise, is to reduce the amount of carbon dioxide emitted when coal burns, either by capturing it before it enters the atmosphere or by treating the coal chemically "to produce a flammable gas with [a] lower carbon content." Fallows examines what is being done in China and the U.S.—mostly in China—to achieve these goals and how the Chinese and Americans are cooperating in the field of energy production.

[179] Scott Allegrucci, "What Might Have Been," *Planet Kansas*, Feb./March 2010.

A matter of days before the Parkinson-Sunflower deal was unveiled in public, a very different kind of announcement was made in Greensburg. This occurred at the grand opening of the John Deere dealership. Dennis McKinney was there—he had just spoken, once again, about the importance of faith. Then Marty Wilkinson, the Senior VP for John Deere Renewables, said that his company would be erecting ten 1.25 megawatt wind turbines approximately three miles southwest of Greensburg. The facility would generate enough electricity to supply the entire town by 2010. "This project will create jobs and reduce greenhouse gas emissions," Wilkinson told his audience. "Additionally, the company's contribution of renewable energy credits (RECs) to Greensburg will help the city achieve its green goal."

With a history of building wind farms—the company had erected 35 over the preceding four years at a cost of $1 billion—John Deere was stepping up to the plate to make Greensburg potentially power-independent in an environmentally sustainable way. One of the wind farm's towers was to be erected in a field belonging to Gordon Unruh, the same Gordon Unruh whose Mennonite family history was recounted in the second chapter of this book.

****

Australians refer to their country as 'Oz.' That suggests a bond between Kansas and the Land Down Under.[180] But there are few places on Earth more remote from the Sunflower State—specifically Kiowa County—than Australia, and the State of Victoria in particular. Victoria, and its capital, Melbourne, are far away. *Far* away. Buried deep in the southern hemisphere, a 14-hour plane ride from Los Angeles, Melbourne is a sophisticated city of more than three million people. Most are descendants of European settlers, but many people have ancestral roots in India. Or China. Or Southeast Asia.

Late in January 2009, there was so much dust in Melbourne's air, the sky turned an ugly brown-grey. This part of the country was enduring a decade of drought and heat. Most lawns were uniformly straw-yellow. Those that weren't had signs which identified these oases in the urban blight as lawns nourished with recycled water. Watering restrictions were enforced everywhere. In the upscale precinct (neighborhood) of Albert Park, gardens were watered before breakfast, and even then, on every other day.

This was the month of the ravenous wild fire that destroyed, among others, the town of Flowerdale, a short way outside Melbourne. It laid

---

[180] Wes Jackson visited Australia in 1997 to engage in a discussion about the role of the ecosystem as a measure for agriculture. It was "the perfect continent to host such a conference," he wrote, because "its soils are generally poor . . . [and] Australia is doomed to stay soil-poor." See Jackson's book, *Consulting the Genius of the Place*, 2010, available on Amazon.

waste vast stretches of eucalyptus forest and vineyards in Victoria. A week before the bush fire began, Chris Delbridge, an employee of the Department of the Environment in the State of Victoria, expressed grave concern over the state of the forest. He believed, like most Victorians, that it was only a matter of time before fire swept through the bush. That may have been what people thought, but no one knew how bad it would be.

February 7, 2009, came to be known in Australia as Black Saturday. The fire that raged to the north and east of Melbourne that day, and the days following, had the power of 1500 atomic bombs of the kind that was dropped on Hiroshima. (Lonnie McCollum, are you reading this?) Enough power, according to a local newspaper, *The Age*, "to power Victoria for a year." *The Age* also quoted a senior lecturer in fire ecology and management at Melbourne University to say that "the best fire-fighting equipment could be used in direct attacks only on fires that burned up to 4000 kilowatts a meter."[181] Some fires on Black Saturday burned at an intensity of 150,000 kilowatts.

One hundred seventy-three people lost their lives in those fires. By some estimates, one quarter of the wildlife in the affected area perished. Chris Delbridge was put in charge of relief operations. He was a very, very busy bloke.

Flowerdale decided to rebuild as green as it could. Community leader Peter Williams said, "The tragedy of Black Saturday has also provided a once-in-a-lifetime opportunity to rebuild in a sustainable way." His town was viewed as the hub-to-be of a region-wide program.[182] Williams found out about Greensburg and its green rebuilding effort from the web. He called Bob Dixson to invite him to spend time in the ruins of Flowerdale. There were similarities, after all, between the two towns and the natural disasters that had wiped them out, most of them tragic but one wry: Both the liquour store in Greensburg and the pub in Flowerdale had escaped while everything else around them was being torn apart or vaporized.

Dixson accepted Williams' invitation. He and his wife Ann visited Flowerdale after Dixson attended an environmental conference in Melbourne— the Green Design and Planning Summit— where the mayor spoke the last week in June 2009. The Dixsons spent two days in Flowerdale, commuting back and forth from Melbourne. Bob Dixson spoke to an audience of 40-50 people about the Greensburg experience. The story of Greensburg was written up in *Habitat Australia*, a magazine published by the Australian Conservation Federation. As a result of this contact, Flowerdale became the first town to consider modeling its reconstruction after Greensburg.[183]

---

[181] "Bush fires like '1500 atomic bombs': Inquiry," *The Age*, May 21, 2009.

[182] This comes from *greencrossaustrlia.org*, the web site of an affiliate of *Green Cross International*, which was founded by Mikhail Gorbachev.

[183] Others would follow. See the last chapter.

For both towns, said Mayor Dixson, "rebuilding isn't about global warming; it's about taking care of what we have and what we are leaving to the future."

A lot of Aussies *are* persuaded that the fires are connected to climate change. A royal commission was set up to find out. A decade of drought has wrought havoc with the country's agriculture. For those who scoff at the connection with rising world temperatures, the *Los Angeles Times* quoted one orchardman to say, "You'd have to have your head in the bloody sand to think otherwise."[184]

Some scientists regard what is happening in the country as a taste of things to come. Again, the *LA Times* quoted Tim Flannery, an Australian paleontologist and an outspoken proponent of the dangers of climate change, as follows:

> Australia is the harbinger of change. . . . The cost to Australia from climate change is going to be greater than for any developed country. We are already starting to see it. It's tearing apart the life-support system that gives us this world.[185]

How much time is left before global warming begins to affect our lives in measurable—and catastrophic—ways? The carbon dioxide concentration in the atmosphere is currently hovering around 390 parts per million (ppm), a level higher than it has been in many millions of years. "We know that the last time [this occurred] much of the Greensland and West Antarctic ice sheets were not there," said the climatologist Michael Mann of Penn State.[186] If the Arctic permafrost melts, this would release enormous amounts of methane since these regions are essentially frozen peat bogs. Methane is a greenhouse gas *par excellence*. A vicious cycle of global warming would be activated by its escape. Is there a tipping point beyond which change is irreversible? There may not be a 'fixed threshold' for such a calamity although scientists look anxiously at the range of 450-500 ppm of carbon dioxide. Mann again: "What we have with rising carbon dioxide levels in general is a dramatically increasing probability of serious and deleterious change in our climate."[187]

---

[184] Julie Cart, "Scientists: Climate change bringing crisis to Australia," *Los Angeles Times*, reprinted in the *Lawrence Journal-World*, April 14, 2009.
[185] *Ibid.*
[186] James Fallows, "Dirty Coal, Clean Future," *The Atlantic*, December 2010.
[187] *Ibid.*

AFTERMATH OF COMPROMISE

> "Then I thought, as the country was so green and beautiful, I would call [it] the Emerald City, and to make the name fit better I put green spectacles on all the people, so that everything they saw was green."
>
> –Wonderful Wizard of Oz

"The world is watching us," Steve Hewitt used to say. Month after month, Greensburg found itself the focus of national and even international attention. The town certainly had the ear of the President, whoever occupied the Oval Office. President Obama, speaking on February 24, 2009, before a joint session of Congress, mentioned Greensburg in his speech:

> I think about. . .a town that was completely destroyed by a tornado, but is being rebuilt by its residents as a global example of how clean energy can power an entire community—how it can bring jobs and businesses to a place where piles of bricks and rubble once lay.

Despite the optimistic note about jobs in Obama's speech, Greensburg was continually frustrated in its search for them. In March 2009, Steve Hewitt expressed the opinion that the biodiesel plant, whose coming to Greensburg had been announced more than one year before, was dead, a victim of the stubborn recession the country found itself in the grips of. Jeanette Siemens, Kiowa County Economic Development Director, gave the same, if slightly less unequivocal, answer to the *Signal* in an interview which appeared on March 13, 2009.

I talked with [VP] Pat Stein [of Torsten Energy] two or three weeks ago and he just said that, with the economy the way it is and the cost of feed stock right now, it's . . .on hold.[188]

Torsten still held the rights to 288 acres of land where it had planned an industrial park northeast of Greensburg. Stein told Siemens that he still hoped to develop the tract as an eco-industrial park.

To compound Greensburg's problems, the assembly plant which CEO Tim Schmidt of Xtreme Structures had come to town the previous July to tout was just as much a will-'o-the-wisp. As previously recounted, Xtreme Structures went belly up in a welter of ill feeling and law suits early in 2009. In hindsight, the ploy of establishing an Xtreme Structures factory in Greensburg appeared like a "Hail Mary Pass," in Hewitt's words. Although the town had put no money into the venture, Hewitt described the loss of the Greensburg plant as "a devastating blow."

The company's bankruptcy was bad news, too, for Greensburg Green-Town, whose chain of eco-homes seemed to be doing a two-steps-forward-one-step backward shuffle. First, the Robert McLaughlin House was on hold for financial reasons. Ditto, the Mother Earth House. In an e-mail message dated March 11, 2009, *Mother Earth News* publisher, Bryan Welch, wrote: "We are still trying to help them build their first green demo house. It has not been easy."

Forget 'first.' That distinction now belonged to another dwelling, the 2100-square-foot silo house by Armour Homes. Located at the intersection of Wisconsin and South Sycamore, the house—basically a concrete cylinder with a garage tacked on—was inspired by the Greensburg grain elevators which had survived the tornado. That spoke to Armour Homes president, David Moffitt, who said:

> After the tornado went through Greensburg, we were impressed that the only thing left standing was the silo, a circular, wind-resistant structure.

To show how strong the silo house was, a Ford Escort was dropped on its roof from a height of 65 feet. The house was undamaged. Not so the car. But it had already been knocked about by the tornado two years earlier. An obvious photo-op, the event was filmed by Planet Green and a host of amateur photographers.[189]

---

[188]Mark Anderson, "Biodiesel plant on the mat for now," *KCS*, Mar. 13, 2009.

[189]See Catherine Hart's article on the Greensburg GreenTown web site, "Car Drop a Smashing Success," Apr. 9, 2009, and the Associated Press article, "Home builder in Greensburg drops car 65 feet onto new home designed to withstand tornadoes," which ran in the *Lawrence Journal-World*, Apr. 10, 2009.

The silo house and the Ford Escort

The $250,000 house became GreenTown's office over the course of 2010. The house also has space for meeting rooms and guest bedrooms to accommodate eco-tourists. There are a number of green features, too—water-saving devices, a roof garden planted by Canadian volunteer students (who left Greensburg studded with Canadian maple leaf flags when they bussed back to Ontario), a bank of photovoltaic cells to generate two kilowatts of electricity—about half the house's needs—and a passive ventilation system.[190] But it would not be LEED platinum-certified. What Greensburg GreenTown said was that certification from the National Association of Home Builders (NAHB) would be sought for the silo house. NREL hoped that that organization's National Green Building Program would be adopted as a standard to complement Greensburg's city code.

The Xtreme Structures bankruptcy cost Greensburg GreenTown a house. Wallach had counted on the California company to partner with K State architecture students working under the direction of their professor, Gary Coates, to put up the second of his houses. A group from K State had actually come out to Greensburg on September 15, 2008, to assess site possibilities for their collaborative effort with Xtreme Structures. As far as its Greensburg endeavors were concerned, K State had rotten luck. The collapse of Xtreme was another blow for the university's department of ar-

---

[190]Some shortcuts on the roof garden were taken, some mistakes made, and all that survived of the garden by October 2009 was a straggly row of cauliflower. A company from Dodge City remedied the situation. Contracted to provide labor and materials to position trays of different-colored sedum on the roof of the Commons Building, the company found that it had sedum left over. The excess was installed on the roof of the Silo House. See the July, 2011, issue of *Greensburg GreenTown*.

chitecture. K State students were more successful when, in mid-December, 2008, fourteen of them came back to Greensburg, with Coates, to make presentations of new work. They brought with them models of sustainable and affordable houses. The models, a booklet with the renderings and plans of the houses, and a CD related to the houses were all left behind as a gift to Greensburg GreenTown.

Daniel Wallach appeared to have scored a coup in securing an eco-home from the University of Colorado, which offered to re-assemble in Greensburg its prize-winning entry in the 2005 Solar Decathlon. This might have been viewed as a replacement for the K State-Xtreme Structures house. But once again—the refrain from a sad song—there was no money to complete the project.[191] The University of Colorado took its solar competition house elsewhere.

In 2009, however, Greensburg GreenTown organized a competition of eco-homes which was won by a firm in New York City. The company, Steven Lerner Studio, submitted renderings for "Meadowlark House," which used a modular wall system called HIB after the German company which developed it. HIB (Highly Insulated Building) involves wood blocks which are assembled like Legos. They resist strong winds—a concern for the gale-buffeted village of Greensburg. Its submission garnered Lerner the grant prize of $10,000 and the chance to see its house constructed.

As with all the eco-homes in Wallach's parade of sustainable houses, construction was contingent on securing the necessary funding. By the fall of 2010, about one-third of the estimated $350,000 to build Meadowlark had been contributed by the Raymond C. and Anna T. Johnson Foundation. Associated with this family foundation was a man called Rob Threlkeld who worked as the manager for green initiatives at General Motors in Detroit. He learned about Greensburg at a conference in San Antonio, Texas, which both he and Bob Dixson attended. A weather buff, Threlkeld later passed a night at the Silo House in Greensburg along with his six-year-old son. They were, in fact, the first guests there. The decision to support Meadowlark House rested, in part, on a coincidence: Threlkeld's Dad had grown up on a street called Meadowlark. Threlkeld "could not believe the connection and decided to provide the seed funding for this house." Shortly afterward, he approached two of the officers of the foundation—one of whom was his Uncle Al—to support the initiative. The contribution of $100,000 was a sizeable one for the new foundation to make. It generally limits its contributions to the $5,000 to $10,000 range although it had given $25,000 for a walkway, dedicated in honor of Threlkeld's father, to the Virginia Museum of Transporation. The elder Threlkeld had worked for the railroads for over thirty years. Although it was shy the full cost of the Meadowlark

---

[191]Wallach said that Greensburg GreenTown would need to find $20,000 to bring the house to Greensburg. In addition, it would take $50 thousand to erect it.

House, Greensburg GreenTown began construction on the home, breaking ground on April 30, 2011. By the fall of that year, the structural skeleton, sheathed in Tyvek, had arisen and was awaiting its windows. A lack of funds postponed their installation; the building was mothballed to protect it against winter weather, 2011-2012.

****

Efforts to promote Greensburg were pursued by Gary Goodman and Mayor Bob Dixson. Two marketing specialists from Dodge City, Bob Wetmore and Mike Wilson, were consulted about regular updates of the Greensburg web page and how these could be done to promote tourism effectively. In March 2009, both men made a presentation before the City Council. Council members were sufficiently impressed to order Dixson and Steve Hewitt to pursue the matter further. Wetmore, a native of northwest Pennsylvania, was eventually hired to be the economic director for Kiowa County, replacing Jeanette Siemens who retired.[192] Wilson's services were retained to market the city's image. Gary Goodman said in early March 2009 that two weeks later there would be a brain-storming session on the same subject with Bill Kurtis, formerly of CBS. This session, which occurred at Kurtis' ranch in southeast Kansas, went well. Goodman said the following month that they worked on the voice-over for the Greensburg web page. Kurtis' improvisation was so good that "if we had captured it on tape, we could have used it as it stood."

Two nights before Obama spoke about Greensburg before Congress, the phone rang in Bob Dixson's home. It was the White House—specifically David Agnew of Intergovernmental Affairs. Dixson was being invited to sit with the First Lady, Michelle Obama, during the President's speech. The next day, in a flurry of e-mails, the mayor was advised to hop a 6:30 p.m. flight from Wichita to Washington, D.C., and to tell no one anything about the purpose of his trip. Arriving at 11:30 p.m., Dixson was invited to attend Tuesday lunch at the White House. There he met and hobnobbed with Secretary of Housing and Urban Development (HUD) Shawn Donovan and Earl V. Delaney, the man who was to chair the Accountability Board for the implementation of the stimulus package recently passed by Congress. Dixson spent the entire afternoon at the White House and then—following the procedure adopted for Steve Hewitt's visit to George W. Bush's last State of the Union address—went by motorcade to the Capitol for the president's speech.

---

[192] At the same time he was given the key to his office in the Business Incubator, Wetmore was given another key to the utilities room of the building and told that he was now the technician-on-site. Since the operations of the Incubator are computer-controlled, the newly-minted economic director had to climb a steep learning curve to master the high-tech heating, cooling, and lighting mechanisms of the Incubator.

The next day the mayor's cell phone never stopped ringing. He became, almost overnight, something of a sustainability superstar, receiving invitations to attend green conferences in locales as diverse as Paris, France, and Melbourne, Australia.[193]

****

Mayor Dixson was carpet-bagging around the planet, playing the role of a Kansas Johnny Appleseed. His reason to go to Australia was clear,[194] but what lay behind his invitation to go to Washington? Were Obama's people keeping up with the Joneses of the previous administration? Bush had shown little interest in environmental sustainability whereas Obama did. So the Greensburg experiment was more in line with the new administration's priorities. Two weeks before the mayor received his call from the White House, Governor Kathleen Sebelius and Secretary of Homeland Security, Janet Napolitano, had presided at the opening of Dillons Supermarket, adjacent to the Qwik Shop, the only place in town where Greensburg residents could buy groceries. Speculation in town had it that Napolitano—a surprise guest for such an event—had come to Kansas to sound out Sebelius about serving in Obama's Cabinet. There was also speculation that the idea of inviting a Greensburg official to the Obama speech arose while Napolitano was visiting the town.

Obama's election, and his speech in February, 2009, influenced some foreign reporters to visit Greensburg. The Danish journalist, Steffen Kretz, introduced two chapters back, wrote in an e-mail that he had chosen:

> to include Greensburg in a series of five news features that all illustrate 'Obama's green USA'—knowing that the folks in Greensburg turned green before Obama got elected—but the story illustrates from my point of view a change in opinion and attitude in the US.

Kretz said that he was surprised "to find progressive green thinkers in the heart of the conservative Bible Belt."

---

[193] Bob Dixson wangled yet another invite to the Obama White House: He attended a reception on December 14, 2009, with 200-250 other invitees. By now, the mayor was on a first-name basis with David Agnew. Dixson talked about the string quartet which played before the President and First Lady descended the staircase from the third to the second floor to greet their guests. He also talked about the hors d'oeuvres, the pecan pralines served for dessert, and the call to vacate the premises after two hours to make room for another presidential reception.

[194] See the previous chapter for Dixson's trip to Flowerdale.

A reporter for the Spanish newspaper, *La Vanguardia*, Marc Bassets, wrote a one-page story about Greensburg after visiting the town. The article begins with Erica Goodman's joke about moving to Greensburg from Las Vegas to seek a more exciting night life—a joke that now had a bilingual life of its own. Like Kretz, Bassets came out to Greensburg as a result of the Obama speech:

> He [Obama] mentioned Greensburg and invited the mayor. My newspaper was working on a series of reports where we tried to portray America in a time of recession, and I found Greensubrg's story compelling.

His impressions of the town?

> The effects of the tornado are still visible [in March 2009]. I was surprised that Main Street was still closed and almost without any building. . . . And, of course, the idea of rebuilding the town in a sustainable way is also very interesting. The fact that most people are conservative and lean to the Republicans, and nevertheless they are building the greenest town in America, defies a lot of clichés and preconceived ideas.[195]

Bob Dixson admitted to Bassets that, even though he was a life-long Republican, he found going green logical.

****

Another dollop of disappointment was to follow the Xtreme Structures bankruptcy. It was hoped that the industrial drought might be broken by Agriboard's rebuilding its factory or a regional branch in Greensburg. But as of February 2010, no decision on this had been made. It was bruited about that the decision was being held up by an insurance issue: How much should Agriboard be reimbursed for the loss of its 'one-of-a-kind' machinery destroyed by fire at its Electra, Texas, plant? The recession of 2008-2010 also hindered the company's decision to expand to Greensburg. As late as the spring of 2010, Steve Hewitt was telling the *Signal* that Agriboard was not dead. "Myself, Mayor Dixson, and Bob Wetmore continue to look for ways to bring this company and its jobs to the county." A full year later, however, the Agriboard logjam remained unbroken, and people ceased to talk of the prospect of scoring this green-industrial *coup*.

---

[195] Marc Bassets, "Renace un pueblo," *La Vanguardia*, Mar. 29, 2009.

Nonetheless, the streetscape, which had been postponed when bids for the work came in higher than the city had budgeted for, was finally underway. March 2009 saw four blocks of Main Street barred to traffic for redesign work. A few changes—more time allowed the contractor for the work and more options—resulted in lower bids a second time round. Whereas initially these had been submitted in the vicinity of $4.5 million, rebids came in around $3.8 million. That, the city could afford.[196] The town was chagrined not to receive any financial assistance for the project from KDOT, which got $35 million from the Obama Administration's stimulus package of early 2009. Steve Hewitt had applied for $1.3 million of that money to rebuild the roadbed, curbs and gutters on Main Street. Kiowa was one of two Kansas counties to be turned down. Fourteen others which submitted applications got something. Had KDOT decided that Greensburg had already sucked up enough state funds? Commented Hewitt:

> I hate to be negative on this because KDOT has been good to us in the past. And I know there's this misplaced belief on a lot of people's part that we have plenty of money here to rebuild. I just hope it's not shared by agencies like them.[197]

Commercial life was slowly returning to Main Street. Auctioneer Scott Brown put together a group of 65 Kiowa County residents, each of whom contributed a minimum of $5,000, to the building of a strip mall. Another of the buildings designed by the Wichita firm of MVP, the mall—which was to be anchored on the south by the old Robinett Building, the only two-story structure in downtown Greensburg to survive the tornado—contained nine retail spaces, each with 1,500 square feet of space.

"We're not wanting to incubate, or graduate people out of our building," said Brown, contrasting the mall with the incubator:

> People can stay in our building as long as they like. I know there are people who would have liked to put their business in the incubator but there just wasn't enough room. Each of our spaces is twice as big as the biggest space in the incubator.[198]

---

[196] Design problems delayed the completion of the Main Street project for months, but the street was finally opened for traffic in late October, 2009. Hitches developed in the automatic watering system for the streetscape. Steve Hewitt later attributed the fault to BNIM which he accused of treating Greensburg as a guinea pig for its innovative schemes.

[197] Mark Anderson, "Hewitt laments KDOT snub of Main Street project," *KCS*, Mar. 27, 2009.

[198] Mark Anderson, "Strip mall now 35 percent bigger," *KCS*, Mar. 4, 2009.

The Goodmans, Gary and Erica, bought the Robinett Building, which, before May 4, 2007, had housed the Centera Bank.[199] Constructed in 1915, the Robinett had served first as a clothing store, then a jewelry store. The Goodmans' plan called for using the ground floor as an antique store to be called *Where'd Ya Get That* and to live upstairs in 1,500 square feet of space. Prior to the tornado, Erica had run another antique store called *Fran's* in a much bigger space. But the height of the ceilings in the Robinett allowed for some creativity in storage and display. As for the space upstairs, "that's enough for Gary and me to live," Erica said. And Alanna? She fell in love with a TV cameraman. In November 2008, she followed him to Los Angeles. Continuing her interest in used clothing, she opened a shop in North Hollywood a bit like her *Snootie Seconds* in Greensburg. By the summer of 2011, however, she was back in Greensburg. More about that in the next chapter.

The Robinett Building

In March 2009, the Goodmans learned that the building they had purchased had been accepted on the National Register of Historic Places, the second such structure to be so designated in town. (The other was the Kiowa County Courthouse, itself then undergoing renovation.) This honor was something of a two-edged sword: Yes, it added to the prestige of the Robinett but it also imposed strict guidelines on how the Goodmans could restore the premises. Still, the designation was clearly worth the trouble it entailed. There were tax credits both from the federal government and the State of Kansas which the Goodmans expected to compensate them

---

[199]Centera Bank began work on its new LEED-certified building, directly opposite the Robinett Building, in the fall of 2009.

for the cost incurred by staying within restoration guidelines. Interestingly enough, it was Lonnie McCollum—who had appeared to be avoiding Greensburg like the plague—it was Lonnie McCollum—who returned, like the prodigal son, to the town where he had thought to spend the rest of his life—to do the electrical wiring for the Goodmans' new store-and-home. *Where'd Ya Get That* opened at the end of October 2009.

Like his father, John Brown, Scott was an auctioneer. It's likely that Scott's son will follow the family calling and become the third-generation Brown to earn his living by auctioneering. John Brown formed the Mullinville Development Corporation to erect a coffee shop where people could come—in Scott's words—"to drink coffee, eat greasy hamburgers, and tell lies." This is what gave him (Scott) the idea to create a 501(c)(3) entity to fund his strip mall. By October 2009, 75 investors had forked over $930,000 for the project, which was quickly taking shape opposite the Business Incubator, on the southeast corner of Main Street and Kansas Avenue (US 54). Among these investors was Denise Unruh's South Central Community Federation which "had given big," to quote Scott Brown. Contributions were like gifts since no dividends would be offered the investors. "If we make money," Brown said, "profits will go to other worthwhile projects in Kiowa County." What seemed to be motivating Brown's fellow investors was a desire to bring retail back to Greensburg.

Although the mall would be energy efficient, it would not aspire to the LEED platinum certification that the incubator had got. Said Brown:

> We've already dug out the concrete of the foundations that were there on our site and ground that up for recycling and the dirt we've used to fill the holes came from Main Street, so you could say that's been recycled as well. But we won't have the money to be as green as the incubator.[200]

There was a question about the mall's falling afoul of the master plan for Greensburg developed by BNIM and adopted by the City Council. There was what Mark Anderson called "an unmistakable contradiction" of the city's master plan. At the City Council's May 18, 2009, meeting, Steve Hewitt said:

> We've won six different awards for our sustainable master plan, and when we get a bit off that plan we [the city leadership] get a little nervous about losing our credibility. Do we need to adjust some things in the plan? I know it's tough to tell people 'no' when a project goes outside the master plan because these are often people we see every day and go to church with.[201]

---

[200] *Ibid.*
[201] Mark Anderson, "Council concerned Master Plan being overlooked," *KCS*, May 22, 2009.

Mayor Dixson was more blunt:

> If we waver on this, we might as well pack up our tent and go home. This community went through a whole lot of work developing this [plan] under the hot dusty tents two years ago.[202]

There were two issues at stake. First, the plan for fifty feet of off-street parking to the north of the mall, bordering US 54. Second, BNIM's Master Plan called for a two-story building on the site of Scott Brown's mall to balance the Business Incubator facing it. With a colic uncharacteristic of a small Midwest town, council members voiced their concern at the deviation from plan. "I did not run for re-election to back off the direction we decided to go," complained Council President Gary Goodheart. Erica Goodman summed up the difficulty by observing that "we've already opened the door a crack and maybe we need to shut it at this point."

Hewitt said that he would pass on the Council's feelings. But the Planning Commission's decision to okay the strip mall's design was not subject to Council approval. The Commission thus was making the decision unilaterally.

Even more fundamental was the issue of Greensburg's continued partnership with BNIM. Hewitt said that he had no problem with BNIM but that he had heard grumbling in the community. Some folks felt that the city should be looking at other architectural firms. One council member questioned why there should be a change when the city "already had a good relationship with BNIM." Erica Goodman, on the other hand, observed that "tourism can be helped by the different designs of various architects. Some cities are known by the variety of architectural designs throughout."[203]

Gary Goodheart said that he could see no reason "to change in the middle of the stream unless we have a real problem with BNIM or their work." That sentiment bolstered—at least temporarily—the Council's consensus to go on working with BNIM. The Kansas City firm was reaping awards—as well as rewards—for its work on the Greensburg Master Plan. In May 2009, BNIM received an Honor Award from The American Society of Landscape Architects (ASLA).[204] The jury for the ASLA award commented that "[BNIM] has created standards that are pragmatic, modest, and achievable, a plan to which other towns should aspire." Stephen Hardy acknowledged the award by saying that "credit for this award and others like it deserved to be shared by the literally hundreds of folks (including

---

[202] *Ibid.*

[203] These remarks are contained in the *Signal* article cited above.

[204] Catherine Hart, "Honors Continue for BNIM Master Plan," *www.GreensburgGreenTown.org*, June 24, 2009.

the great people at GreenTown) who contributed to the process." But locally the plan which BNIM put together was generating more static than praise.[205] A poll taken by the *Signal* in the summer of 2009 found that more than half of the respondents felt that the Master Plan could benefit from major revisions. Be that as it may, a meeting during the summer of 2009 between the City Council and the Planning Commission left the BNIM Master Plan intact. But Scott Brown's parking lot stayed where Brown wanted it.

****

One clear-cut success story from the private sector came from the Estes brothers, who rebuilt the John Deere distributorship to meet LEED platinum certification. Shortly after the tornado, the brothers had met to talk over the situation with Scott Brown in Mullenville, at the very restaurant Brown's father had built for locals "to drink coffee, eat greasy hamburgers, and tell lies." That led to a meeting in Brown's Greensburg office attended by approximately 170 people. There it was decided not to abandon the town but to rebuild. As related earlier, a dispute over building regulations led to the stormy Council meeting at which Kelly's harangue proved to be the straw that broke Lonnie McCollum's resolve and led to his resignation as mayor. But the Estes brothers later made a fateful decision to rebuild green. How did that happen?

According to Mike Estes, Daniel Wallach proved persuasive in urging them to go green. Governor Sebelius also pitched in with her support of the idea, and the city adopted a green standard in December 2007. The Esteses had moved out of town by then, acquiring 77 acres by the airport. But they decided to support Greensburg by shooting for the top: LEED platinum certification. John Deere threw its weight behind the move, and the Esteses hired BNIM to do their LEED application. Mike Estes exhibited a scorecard on which the brothers kept track of points for LEED certification. According to the company's tally, John Deere would come in with 54 points; 52 were required at that time for platinum. Acquiring those points was no pushover. To do it, the Estes' architect, *inter alia*:

- provided in-floor heating in the vestibule;
- installed a burner to use up waste oil from the repair shop;
- erected a 5-KW wind turbine to provide 15 percent of the facility's energy needs;
- equipped counters with LEED-certified renewable wood surfaces.

---

[205] Problems which developed with BNIM's streetscape design, with its automatic watering system, were most easily alleviated by the manual watering of plants along Main Street.

BTI's ambitious plans cost the Esteses an additional $300,000. The new facility opened in January 2009, but the grand opening occurred four months later, two days before the second anniversary of the tornado. In explaining his company's positive reaction to the notion of going green, John Deere Manager Dave Jeffers said that:

> First, the Greensburg facility represents the John Deere image. Second, there was the energy-saving technology incorporated in that facility, and third, there was an overall design—simple, basic, functional, and flexible—which appealed to us.

On May 2, 2009, a large crowd of townsfolk and local dignitaries gathered in the hangar-like workshop of BTI John Deere for the grand opening of the facility. Who can say what lured them in: pride of accomplishment? curiosity? the promise of a buffet lunch at the end of the program? A line of BTI employees stood to the right of the audience, like soldiers at parade rest, dressed not in Army fatigues, but in the pale green polo shirts of the John Deere uniform. After a raft of congratulatory speeches, the Estes brothers and some of their employees hoisted the brothers' mother, Wanda, onto the improvised stage. Wanda Estes was 85 years old and confined to a wheelchair. Once in place, the curly-haired octogenarian cut a green-and-white ribbon with an enormous pair of scissors. The crowd applauded, a gospel band struck up a medley of gospel songs, and people filed out to a tent where plates of barbeque had been laid out on a pair of trestle tables.

One person in the audience that morning was Lynn Billman, who took time to share her glee with some of the locals. The National Renewable Energy Laboratory(NREL) had experienced a ten-fold increase in its budget for renewable energy since the Democrats had taken control of Congress in the 2006 election. She distributed a folder of DOE releases including one specific to Greensburg and another to BTI John Deere which now prided itself on being "the greener dealer." The brochure about Greensburg mentioned the wind farm to be built south of town with its promise of "100 percent renewable electricity, 100 percent of the time." Henceforth the town would "not use electricity generated from fossil fuels, such as coal." Greensburg, the pamphlet announced, was "the first city in the world to adopt these kinds of [sustainability] resolutions."

Ground was broken for the John Deere wind farm on October 23, 2009. The event was the salient go-green achievement of the season. It had come about after Tom Vilsack, Obama's Secretary of Agriculture, announced earlier in the month that the Feds were offering a loan of $1.74 million to Wind Farm LLC, a subsidy of John Deere. The Greensburg farm would cost $23.3 million. The difference would be made up by Deere & Co. as "equity investment." It had rained for two days before the ceremony, and there was some concern whether (unpaved) road conditions would throw a spanner into the agenda. People were instructed to drive out to the Mennonite Church—this was Bethel, the church where Lloyd Goossen and Gordon Unruh worshipped—to be bussed to the wind farm site a mile further south. The wind did not miss its cue, gusting up to 35 mph, that Friday afternoon. Bob Wetmore reckoned that the crowd showing up was 200 strong. That included representatives from USDA and NREL and, of course—weren't they ubiquitous?—a camera crew from Planet Green, which began its third season of filming in Greensburg at the ground-breaking ceremony. Bob Dixson took a historical view of things:

> As our ancestors were here, they knew about wind. They knew about solar when they settled this country. . . . And now we're going to see wind that's going to generate electricity for us. We're going to be the new pioneers of the 21st century.[206]

Yet Greensburg would not be getting its energy directly from the wind farm. The *Dodge City Globe* reported that:

> The city will receive rights to renewable energy credits from about one-third of the wind farm. The Vermont-based company, *Native*Energy, which provided some of the up-front financing for the project, will buy the remaining credits and convert them to carbon offsets for its customers.
>
> . . . . . .
>
> The Kansas Power Pool (KPP), a municipal energy agency that includes Greensburg, is buying power from the project under a long-term purchasing agreement.[207]

The KPP was formed in 2005 when member cities were given notice by the Southwest Power Pool (SPP) that their long-time supply contracts were being cancelled. Working collectively, these cities were able to reduce their costs substantially.[208]

Greensburg GreenTown estimated that the wind farm would be operational by March 2010, and it was. Bob Wetmore reported that the wind turbines were merrily spinning away in late February 2010.

---

[206] Eric Swanson, *Dodge City Daily Globe*, Oct. 24, 2009.
[207] *Ibid.*
[208] *www.kansaspowerpool.org*

In August 2011, KPP proposed that Greensburg, along with other municipalities participating in the pool, join with KPP to buy a natural-gas burning power plant in Missouri. The purchase, it was contended, would help to satisfy participants' growing energy needs. Natural gas had become less expensive because of a process known as hydraulic fracturing ("fracking"). But there was evidence that fracking caused environmental damage,[209] and—as pointed out in the *Signal*—"purchasing a natural gas plant may become a moral issue for a city promoted as a leader in renewable energy consumption."[210] Be that as it may, Greensburg agreed to the KPP proposal.

****

A week before the grand unveiling of BTI John Deere, the Sunchips Business Incubator opened for public viewing. Angee Morgan of Kansas Emergency Management came back for the event which drew a smallish crowd. Surprisingly, there were fewer celebrities than for the opening of Dillons Kwik Shop earlier that year. But Craig Piligian, president of *Pilgrim Films & Television* and a major financial contributor to the project, was present. So were a few people from Planet Green. Students, given the afternoon off, hawked free packages of Frito-Lay. Steve Hewitt started things off by calling the incubator "the smartest building in the state of Kansas." An alarmingly sunburned Bob Dixson stood head and shoulders above the cluster of people milling about the main entrance to the building. After the speeches, people were allowed to tour the building. The facility already housed a studio operated by Scott and Susan Reinecke. The Reineckes' store, named Studio 54 after the date of the tornado—like the Art Center two blocks away—was selling stained glass pendants and knickknacks which looked incongruously Victorian in their 21st-century surroundings. Among the items for sale were pieces salvaged from glass shattered by the tornado. In October 2009, Susan announced that the store had received a commission to do three 14-foot stained glass windows for the sanctuary of the reconstructed Methodist church, a part of the building that was slated to open a couple of months later. The incubator also sheltered a cosmetics store, with a massage table curtained-off for privacy, and an environmentally friendly paint store.

The business incubator was LEED platinum-certified, but other building projects had to scale back their sustainability goals. The Shanks' GM distributorship, for example, had green elements in its new building—light shafts, superior insulation, and efficient heating/cooling—but short of what LEED required for platinum certification. Chevrolet had helped to fund

---

[209] Joe Spease, "Why Shale Gas from Hydraulic Fracturing Is Not the Clean Energy Solution We Need," *Planet Kansas*, Auguyst/September 2011.

[210] Patrick Clement, "Power Pool Pitches Proposed Project," *KCS*, Aug. 10, 2011.

Inauguration of the Business Incubator

the new building, and the auto company used the Shanks' dealership to introduce its plug-in hybrid, the Chevrolet Volt, which came on line in 2010. Going green, which was more in keeping with the spirit of the evolving US auto industry, may have saved the dealership at a time when a bankrupt General Motors was axing dealerships from coast to coast. In October, 2009 the Shanks received a 'go forward' letter rather than instructions to 'wind up.'

The new First Baptist Church, another project with LEED aspirations, was built on Kansas Avenue (US 54) opposite the Kwik Shop. It was decided ultimately not to apply for LEED certification because of the expense involved. An unfinished parking lot beside the church—after all the rain the region endured the week before the second anniversary, it was a strip of shoe-sucking mud—bore witness to the shortage of funds needed to finish up. "We are not so strong on environment," Pastor Marvin George admitted the day after the inauguration of the church, "but we hit energy hard." This was seconded by the architect for the building, Kelly McMurphy, who said that ICF blocks had been used in the foundation. The church was "super-insulated," and energy efficient appliances, lighting, and HVAC had all been installed.

****

Greensburg was unique in the US. But major green developments abounded elsewhere. There was, for example, a race among several nations to be the first carbon-neutral on earth. The US, sadly, was not among them. The favorite (in 2009) was Costa Rica which aimed at going carbon-neutral by 2021.[211] By the end of the first decade of the twenty-first century, the country was already generating 80 percent of its energy from renewable sources. New Zealand and Norway were also in the running to be the first "carbon-free" countries on earth, but Costa Rica planned to beat them both by a margin of 20 to 30 years.[212]

China—much criticized for its use of pollution-generating coal—was touting plans to erect cities with populations of half a million which would get their energy from human waste, the wind, the sun, and biofuel. One city in northeastern China—Tianjin—became home to five wind turbine manufacturing plants erected by the Danish company, Vestas. The People's Republic is now the world's largest maker of wind turbines.[213]

Wind, solar and biomass were slated to supply eight percent of Chinese electricity-generating capacity by 2020, but coal would still account that year for two-thirds of the country's capacity.[214]

Then there was Masdar, the fabled city of sustainability, car-free and wind-turbined, with PV panels on every corner, the eco-center of Abu Dhabi, one of the constituent states of the oil-soaked United Arab Emirates.[215] Laid out by the British architect, Norman Foster, the city would be carbon-neutral and garbage-free thanks to ambitious recycling plans. Combustion-engine vehicles were to be banished from the city, to be replaced by a fleet of computer-controlled electric cars which ran, not through the city but underneath it. A comparison between Masdar and Disneyland did not escape Norman Foster. He noted that, like the amusement park, Masdar buried its services underground. At an estimated cost of $22 billion, the town was projected to house 90,000 people by the year 2016. The concept dwarfs what was happening in Greensburg. But its realization has brought some astringent criticism. Can Masdar "ever attain the richness and texture of a real city," wondered Nicolai Ouroussoff, architecture critic for the *New York Times*.[216] Ouroussoff toured the town after the first phase

---

[211] Frank Bures, "Green Acres," *Wired*, Feb. 2008.

[212] "Maldives joins Costa Rica & New Zealand in Race to go Carbon Neutral," www.evolvingchoice.com, Mar. 16, 2009. See also Alana Herro, "Costa Rica Aims to Become First 'Carbon Neutral' Country," www.worldwatch.org, Mar. 12, 2007.

[213] Keith Bradsher, "China Leading Global Race to Make Clean Energy," *NYT*, Jan. 30, 2010.

[214] See the preceding chapter for comment on China's development of clean coal.

[215] Grégoire Allix, "Des tournesols géants au coeur de la future écocité de Masdar," *Le Monde*, Sep. 5, 2009. Near-by Qatar hosted a conference on the *Humanization of Cities of Tomorrow*, Oct. 4-5, 2011, to which Bob Dixson was invited as a VIP speaker.

[216] Nicolai Ouroussoff, "In Arabian Desert, a Sustainable City Rises," *NYT*, Sep. 26, 2010.

of construction was completed. While admiring Foster's "inspired synthesis of ancient and new technologies," he goes on to compare Masdar with gated communities in which "both the megarich and the educated middle classes have increasingly found solace by walling themselves off inside a variety of mini-utopias."

Tianjin and Masdar were recognized among four "eco-cities" by environmental writer Ben Jervey, who assessed them on a scale from 1 to 10.[217] Masdar, which got a rating of 8 for the likelihood that it would live up to the hype that it had generated, and Tianjin, with a rating of 5, came in second and fourth respectively in Jervey's survey. A solar-powered town in Florida, Babcock Ranch, came in third. First place went to none other than Greensburg (rating: 9). To earn such high marks, Greensburg was seen as something of an anomaly in its part of the country. Said Jervey:

> Given all the national attention focused on Greensburg, it's nearly impossible to imagine anything short of great things. Leonardo DiCaprio has produced a show about the town's plight for Discovery's Planet Green and the respected carbon offset company *Native*Energy has created a special offset product for the town. And while there won't be too many lessons for cities to learn from the small town's eco-resurrection, Greensburg will certainly live up to very lofty expectations and prove the benefits of sustainable design at any scale.

Greensburg thus stood in contrast to other projects in the world which had *not* lived up to expectations. Jervey cited, in particular, a much-hyped suburb of Shanghai which was supposed to have provided an environmentally friendly home for half a million people. Plans for the town never left the drafting board. Blame was hard to ascribe, but Jervey noted that the official who was put in charge of the project was now serving 18 years in jail for fraud.

---

[217] Ben Jervey, "The Death and Life of Model 'Eco-cities,'" *www.good.is*, June 8, 2009. *Good*, an integrated-media platform, exists, according to its web page, "for people who want to live well and do good."

CAN THE CENTER HOLD?

> *"Hush, my dear," he said. "don't speak so loud, or you will be overheard—and I should be ruined. I'm supposed to be a great Wizard."*
>
> *"And aren't you?" she asked.*
>
> *"Not a bit of it, my dear; I'm just a common man."*
>
> –Wonderful Wizard of Oz

By the fall of 2011, some things had happened; other things were ongoing. One loss to the community was the closure of the Lunch Box in December 2009. This was the diner where Megan Gardiner had worked the night of the tornado. Its owner decided to close down and move her business to Mullinville where she renovated and reopened the Mullinville café where, according to Scott Brown, people came to eat greasy hamburgers and tell lies while they drank their coffee.

Another loss—a more tragic one—was that of Pat Jones, who succumbed on February 13, 2011, to what was described as anoxic brain injury. Charlie explained that Pat had suffered from this problem five years previously. It was a condition in which one can breathe in but not out. The call which the Joneses' son, Jesse, placed to 911 did not save Pat. After her death, Charlie and Jesse parked outside their house a trailer with a sign 'Cans for Pat' affixed to its wire mesh covering. By late March, it was pretty much full. Charlie confided that he would contribute $50 per load to the American Lung Association.

To take his mind off Pat's death, Charlie took to tinkering with an array of objects which he had unearthed from 85 loads of dirt scooped up from the intersection of US 54 and Main Street. The dirt had been dumped, at Charlie's request, on the Joneses' property. His findings included mule shoes, coins from the 1820s, a printing press shim dated 1878, and the bowl of a reed pipe, 1820-1840. Charlie was clearly psyched by what he had unearthed. The Historical Society of Kiowa County expressed interest in Charlie's treasures, but he was unwilling to part with them when told that he could not bring them back home from time to time.

A FEMA inspection team which returned to Greensburg in July 2009 to check the town's recovery efforts was impressed. Said one member of the team:

> It's remarkable, the progress, the vision of the community and how it's been implemented. We toured the incubator, BTI and went through the Courthouse as well as looked over the streetscape, and all are eye-popping.[218]

Resistance to going green ebbed and flowed, never going away, lingering like a leitmotif—a *Schadenfreude* which surfaced at each setback encountered by the City Council or Kiowa County officials. Was Greensburg viewed by its neighbors the way Masdar was viewed by Nicolai Ouroussoff, as a town "lifted on a pedestal and outside the reach of the world's citizens"?[219]

On March 3, 2010, the city Planning Commission gave up its effort—"meekly" in the opinion of Mark Anderson—to extend Master Plan criteria to a potential building site east of Greensburg. The announcement was received with applause from the audience attending the meeting.

The Big Well Museum was another victim of residents' exasperation with the cost of rebuilding, specifically rebuilding green. Months of nitpicking ensued after BNIM unveiled its plans for the structure. Even after being pared down to half its original size, it would cost $3 million for 6,500 square feet of space. Townsfolk sniped at the space allocated the gift shop, the number of toilets in the building, even the design of a covered walkway leading into the building. The wall holes in the BNIM design reminded some of Swiss cheese. Chopping out what Greensburgers called "the pretties" was first on the City Council's agenda, although these items had their supporters. Levi Smith offered an opinion that, as a resident who would be around for years to come, he could appreciate the so-called

---

[218] Mark Anderson, "FEMA team returns to review progress," *KCS*, July 16, 2009.

[219] When interviewed in mid-2011, Steve Hewitt expressed his amazement at the resentment he encountered in the neighboring communities of Haviland and Mullinville on the subject of Greensburg's sustainable rebuilding.

pretties. The cost of going green, however, clearly stuck in the craw of other residents. Sustainability was not the problem, contended BNIM's Stephen Hardy. The markup, he claimed, came "completely from the remoteness of the site." His argument did not prove persuasive. Residents thought that, just for once, the town could forego its ecological aspirations.

The controversy over the design and building of the museum illustrates what one county official described as the fatigue people were feeling after three years of going green and reading about the cost of doing so. Resources, both human and financial, were wearing thin. Mayor Dixson had this way of putting it:

> We're a very tender community. . . right now. We've been running on adrenaline for three years, and now our emotions are catching up with us. Now we are back to so-called normalcy.[220]

The case of the Big Well Museum illustrates how the tension built up. Sentiment wobbled all over the landscape of discourse in the first half of 2010. At the meeting on January 25, 2010, a BNIM architect presented the firm's revamped design for the building, one inspired by the well itself: a cylinder partially subterranean and concentric about the well itself. No one quibbled with the poetry of the structure design. But it was Erica Goodman who wanted to know about the adequacy of the restrooms. And the gift shop. The toilets met code, BNIM said. There was always the budget to consider, the financial constraints on what the community could have in its new facility. Bob Dixson wanted to know about the project time line and whether the project would jeopardize the well itself. PEC engineers had been consulted about the question of the well's stability. They had expressed confidence that this would not be an issue. "A circle is inherently stable," one of the BNIM team assured the Council. Scott Brown wondered about the number of potential shoppers in the gift shop at any one time. And what about a concessions stand? Maybe eliminating some of the building's "pretties" would open up more space. Stacy Barnes, director of the project, said that she was always being asked whether visitors could descend to the bottom of the well, something prohibited by code unless safety issues could be successfully addressed.

Late in the summer of 2010, the City Council voted unanimously to reject the bids that had been submitted to build the BNIM cylindrical museum. The lowest of these came in $400,000 above the $3 million target price set for the project. Could the project be shrunk down to get bids at or under the projected price? Scrapping the geothermal heating/cooling

---

[220] Aaron Barnhart, "Three years after tornado, renewal hasn't come easy in Greensburg," *Kansas City Star*, May 2, 2010.

system and a covered walkway might do the trick but at the sacrifice of LEED platinum certification. Steve Hewitt told the *Signal*:

> The Council can now explore some options . . . If they don't like any of those they can always go back to the original design, re-bid it and see if it comes in close to the original figure, or maybe even a little lower. [221]

But, Hewitt cautioned, a different design would put the completion date for the project off by a year.

A follow-up meeting, attended by Bob Berkebile himself—the 'B' of BNIM—on Aug. 16, 2010, hashed over some old as well as some new issues. A majority clearly wanted the project to stay on budget. Some people took issue with the BNIM design. They ridiculed it as an "oil drum," an inappropriate shape for a building to house the Big Well. Was it humanly possible to satisfy the demand for change while achieving LEED platinum certification? The *coup de grace* came a week later. Voting 4-to-1, the City Council instructed Steve Hewitt to "pursue a new direction" on the design of the Big Well Museum. This was reported in an article which began, under a strikingly timid title, on the lower left corner of the front page of the August 25, 2010 edition of the *Kiowa County Signal*. The article continued in full spate on the following page. The Council meeting had followed a 3-hour public forum held the previous day. Again Berkebile, accompanied by Ruth Wedel, was in attendance at the Council meeting. After Wedel listed what seemed to her to be the six priorities to emerge at the forum, Berkebile took the floor to draw an analogy between what Greensburg was struggling with and an anecdote about Walt Disney. Disney had been urged to survey the general public to gauge its tastes before opening his theme park in Anaheim, California–the same Disneyland to which Norman Foster had compared Masdar. The movie mogul shrugged off this recommendation with the remark that "What I have in mind they've never seen."

"There are some decisions the public can't make for you," Berkebile advised the Council. "You can't have the area and all the exhibitry we've laid out in the previous design with this budget if we or someone else redesigns this."

After the meeting, Hewitt said that the Council would need to decide whether to retain BNIM as the project architect and McCownGordon as the construction manager.

Mark Anderson editorialized on the way things were failing to come together:

---

[221] Mark Anderson, "Council drops bids on round museum design for Big Well," *KCS*, Aug. 4, 2010.

There would likely have been no sudden urge to again consult public opinion on the design if the bids had not come in higher than expected. Suddenly the indecisive needed political cover, and they sought it by first pointing fingers at BNIM.

. . .

Out of a population of 800 barely four dozen folks were at the electronic voting session with BNIM on Aug 15, most if not all being over the age of 50. The only person under 20 I recall speaking up recently at a public meeting on the matter . . . told the council his architectural buddies at KSU had raved over BNIM's original design.[222]

At a September 7 meeting of the Council, it was unanimously agreed that McGownGordon be retained as construction managers for the Big Well Museum project, but the Council split, 3-2 (Erica Goodman and Rex Butler—both recently re-elected to the Council—cast the two dissenting votes),[223] on offering BNIM a new contract. At its meeting on September 27, 2010, the Council voted unanimously to open up the design of the museum to other architectural firms. Rather than going the route of a design contest, the Council opted to approach firms via an RFP (request for proposal). This discouraged Bob Berkebile from participating. He said that BNIM had no problem in competing with other firms, but he was disappointed that Greensburg would be asking simply for an RFP. This required a less robust definition of expectations than a design competition. He said: "If it's clear the Council has lost confidence in us, it's best for us to bow out."[224] Berkebile cautioned the Council "to think through what you're asking of those other firms." He added: "Don't forget that we had public input and worked with the advisory board and then late in the process you asked us to open up to public comment again."

This switcheroo had an indirect cost for Greensburg. As the *Signal* pointed out, the $400,000 spent thus far on design and travel—money which had come from FEMA and USDA—was currency under the bridge. Neither

---

[222]Mark Anderson, "UnreMARKable Reflections," *KCS*, Aug. 25, 2010.

[223]They were re-elected on April 1, 2010. Erica received far more votes than any other of the six candidates. Rex Butler returned to the City Council after a hiatus of four years. He was known to be critical of Steve Hewitt. One of the defeated candidates for the Council was John Janssen, who lost his bid to re-engage in Greensburg's civic life. He was seen so little around town after the election, holed up in his office on his farm in Edwards County to the northwest of Greensburg, that one Greensburger described him as a recluse.

[224]Mark Anderson, "Council votes to look past BNIM for Well design," *KCS*, Sept. 29, 2010.

agency would pony up for more design costs. Those would have to be paid by the city itself.[225]

The nod for the new design of the museum went to a Wichita-based firm, Law Kingdon, which submitted renderings of a 5900-square-foot cylindrical building, perhaps a bit less oil-drum-y than what BNIM had proposed. Again the public was invited to comment on the design of the museum at a meeting in late January 2011. Naturally a major concern was that the project stay on budget. Meaning that the whole enchilada—excluding architects' and engineers' fees—come in at or below $3 million. The RFP which had gone out in October 2010 expressed the hope that the museum exhibits, price-tagged at $750,000, be included in the construction costs. Law Kingdon had hoped to keep this expense separate. To stay on budget, the company floated several options to reduce costs. One of these was to shoot for a lower level of LEED certification. The City Council said that it would consider this only as a last resort. The Council was amenable, however, to axing the green roof which had figured in the original plans for the building, an extra which would have cost about $150,000. Its elimination would entail the sacrifice of not a single LEED point, according to Law Kingdon.

There was an evolution of opinion over the summer of 2011 about certification for the Big Well Museum. On June 29 of that year, the *Signal* reported that the City Council was considering options which would lower the price of the building, but "in so doing the Platinum LEED Award would be in jeopardy." On August 17, 2011, the *Signal* reported that the Council had decided to remove the geothermal HVAC unit from the project, saving $200,000 but losing any possibility of LEED certification. "We are trying to balance environmental sustainability with financial sustainability," commented Bob Dixson. The cost of the museum had been brought down to $3.2 million. On August 22, 2011, the City Council voted unanimously to approve the reduced Big Well Museum design. Did the Council's decision violate the 2007 ordinance requiring all public buildings with more than 4,000 square feet to conform to the platinum level certification? No, according to the city lawyer, since the ordinance allowed some wiggle room. The ultimate design for the building fell short of the points required for certification. Budgetary considerations thus trumped environmental goals.

The decision to forego LEED certification did not sit well with everyone. "Not going for LEED certification sucks," opined ex-mayor John Janssen. The building would become a focus of civic life in Greensburg. It needed to be green. "It broke my heart," said Steve Hewitt, who considered the Council's decision to be a huge mistake. In his opinion, the town had lost

---

[225] Mark Anderson, "In search of a design," *KCS*, Dec. 15, 2010.

credibilitty by opting out of the certification process, thereby veering away from the original vision of going resolutely green.[226]

There was some support for the Council's abandonment of LEED certification for the last of the public buildings to go up after the tornado. Look at it this way, said one local resident: Greensburg may have set the bar too high in shooting for platinum certification, so that achieving something less—as praiseworthy as that something might be—looks like a failure.

There was grumbling as well about the Commons Building, a county project closely identified with Gene West. On November 4, 2009, the *Signal* ran a letter to the editor from a Haviland man complaining that the Media Center in the Commons was unnecessary.[227] A petition drive to bring to a vote a $600,000 revenue bond designed to finance part of the Commons was circulated. It quickly acquired the signatures needed to force a vote on the issue.[228] Was the impetus for the petition coming from outside Greensburg? Gene West told the *Signal* that it was. "Demographics and geography" were involved in the conflict.[229] A transfer of funds short-circuited the petition drive: two county commissioners—Gene West and one of his colleagues—voted to transfer $100,000 from a county equipment fund and $450,000 from a surplus in the general fund at the end of 2009. The third commissioner voiced a concern of the Haviland dissidents, that the media center would fail to be financially self-sustaining. But West's ploy worked: the Commons went out to bid on January 20, 2010, one day after the County Commission had signed a letter of condition with USDA for a $1.66 million grant. His determination to see the Commons Building through to completion probably cost West his bid for re-election to the Board of County Commissioners in the fall of 2010.

****

The Twilight Theatre reconstruction project offered a final example of the difficulties encountered in rebuilding Greensburg. A concert to raise some cash for the Twilight was a fiasco, earning zip for the project. Even performers in the country-'n'-western show had a hard time getting paid. A

---

[226] Hewitt made these remarks in September 2011, by which time he had left his position as City Manager of Greensburg.

[227] The same Haviland man later wrote another letter to the editor (Feb. 3, 2010) commending Scott Brown on "his perseverance and wisdom in putting together [his mall] . . . for a mere pittance compared to some of the rest of the construction." The writer went on to assert that the building "is very pleasing to the eye and fits well with what Main Street used to look like." One architect intimately involved in developments in Greensburg, 2009-2011, observed ruefully that people were reverting to a pre-tornado mentality.

[228] More than 60 percent of the 116 signatures obtained were those of Haviland residents.

[229] Mark Anderson, "Is Haviland petition a threat to Commons project," *KCS*, Nov. 4, 2009.

meeting on June 10, 2010, with the executive director of a historic theatre in nearby Hutchison revealed some embarrassing gaps in the planning for the Twilight. There was no rental policy in place for special events like weddings or private parties. Nor had any provision been made to interchange sets or lighting elements for plays or musicals. Adding a system to do that would probably cost an additional quarter million dollars. That, at least, was the estimate of John Janssen, treasurer of the Twilight Theatre board. Plus another $50,000 for an architect since the firm that had done the design for the new Twilight—Spangenburg of Wichita—did not have the appropriate experience for such a system. It was suggested more than once that a fund raiser be hired. And what about annual memberships for the Twilight? The current 100 or so members had paid $100 back in 1989. Gary Goodman was busy selling memberships to the Twilight up to the July 9, 2010 meeting of the theatre board, a meeting notable for the replacement of four of the five board members.[230] Only Scott Eller survived the purge. Gone were Chair Farrell Allison and Treasurer John Janssen. Gary Goodman, perhaps because of his hustling to sell theatre memberships, was elected to the board.[231] So was Matt Deighton. What precipitated the wholesale turnover was a conversation between Goodman and theatre tycoon Bill Warren, who had committed to equip the new theatre once it was built. Warren withdrew his offer of 500 new seats, offering instead 500 used but relatively new ones, which Goodman accepted.[232] To honor his commitment to provide the sound/projection system for the Twilight, Warren laid down three conditions:

- Replacing the board;

- Setting a date for construction to begin no later than May 1, 2011, and to end within a year;

- Competitive bidding for construction.

The Warren ultimatum, as stated by Goodman, contrasted starkly with what both Allison and Janssen had recently reported about the movie man's commitment to the Twilight Theatre project.[233] In an opinion piece which ran in the July 14, 2010, issue of the *Signal*, Anderson wrote that "With a

---

[230] Mark Anderson, "Twilight board reborn," *KCS*, July 14, 2010.

[231] Gary Goodman's political life had been put on hold—at least for a few months—when he was fired from the city's Planning Commission by a 3-0 vote of the City Council on March 1, 2010. The roots of this turn of events lay in antagonism between Goodman and Steve Hewitt, whom Goodman had accused of lying. Goodman may have been down, but he was not out. It was his wife Erica who had advised him to become active in selling memberships for the Twilight Theatre.

[232] The day before Goodman's interview in October 2010, he and Rex Butler had trucked out to Wichita to take possession of Warren's chairs.

[233] Mark Anderson, "Twilight board reborn," *KCS*, July 14, 2010.

guy like Warren and what he's offering to bring to the table, you do what you can as quickly as you can to convince him matters are moving forward, especially when you quote Warren as saying he's got a 'large sum of money from other people' if the project gets done by May 12, 2012."

The new board gave signs of moving in the direction Warren wanted. Three days after the shake-up of July 9, 2010, the board talked about raising $18,000 by mid-August to qualify for a matching grant from the South Central Community Foundation. It agreed on the need for a special fundraiser.

Things got off to a good start: a gift of $5,000 from the local Lions Club put the Twilight over the top to garner the $100,000 matching grant from South Central. And Spangenburg redid the theatre plans, chopping ten feet off its depth, in an effort to bring down the cost of the building.[234] The new Board of Directors for the Twilight began to explore links with the Media Center in the Commons Building and with the Kiowa County school district. Children at the school were recruited to raise money for the Twilight. Students were able to offer the Board a Christmas gift of $1349.06. The school shop pitched in by selling stepstools. "The small ones are the perfect height for a child to step up to the sink to brush his teeth," said the shop supervisor. "The bigger one is the perfect height for an adult to reach that thing which is on the top shelf."[235] $20 for a small stool, $25 for a large one. The possibility of a link-up with the school and the Media Center got Bill Warren's attention. He ordered the architect, Ron Spangenburg, to redesign the theatre double-time so that the Board could view the alterations at their next meeting. Plans called for hardwiring the theatre to broadcast live events at the Twilight via the Media Center. "We could put up the building right now if we needed to," said Gary Goodman. "But, of course, with a theatre we need a lot to put inside it."

****

What happened in Greensburg *happened* only because of a handful of people with the vision and the determination to pull off a stunning change in the way private citizens and institutions built; how they preserved their resources; and how they developed sustainable sources of power. What people like Steve Hewitt and Daniel Wallach accomplished "ravishes into admiration," to steal an oft-quoted phrase by David Hume. In 2008, Steve Hewitt was recognized by *American City and County Magazine* as the Municipal Leader of the Year. The award was made, according to the magazine, for "someone who has vision beyond their leadership and courage beyond the

---

[234] See Mark Anderson's interview with board chairman Chuck Miller, in the Nov. 10, 2010, issue of the *KCS*.

[235] Kim McMurry, "Twightlight Theatre Time. . .", *KCS*, Jan. 19, 2011.

Architect's rendering of the new Twilight Theatre

officialness of their title." Nominated by PEC's Tim Lenz, Hewitt was featured on the cover of the magazine under the sobriquet "Come-back Kid," which figured in the article the magazine ran about him:

> Determined not only to salvage his hometown, but to recreate a model of energy efficiency and sustainability, Hewitt never considered giving up. [236]

A second award came in November 2009, when *Governing* magazine included Hewitt in its roster of state and local officials to be honored for leadership and excellence. This award, unlike the previous one—which recognized the success of the immediate response to the tornado—focused on long-term recovery.

Despite the national recognition of his stick-to-it-ness to see the job done, Steve Hewitt *did* throw in the towel and walk away from Greensburg. The first temptation to do so arose in early 2009. He was approached by a consulting firm undertaking a search for a Wichita suburb, Valley Center, to fill that city's vacant administrator position. Negotiations on Hewitt's applying for the job were kept secret until it was revealed in the Wichita media that an interview was upcoming. At that point, Hewitt contacted the *Signal* to confirm that he would, in fact, be interviewing for that position. "I didn't plan on going public on this now," he told Mark Anderson, "but [Wichita TV stations] got hold of it, and I can't deny it." He went on to say:

---

[236] Nancy Mann Jackson, "Municipal Leader of the Year: Come-back kid," *American City and County*, Nov. 1, 2008.

> I don't really want to leave at this time, but with the way some [things have gone lately], I feel I need to go listen to them and hear what they have to say.[237]

Bob Dixson told the *Signal* that he was "taken by surprise" at the announcement of Hewitt's candidacy for the position in Valley Center. The mayor said:

> Steven (*sic*) has done us an excellent job. [His applying] is something that's in the progression of things. If that works out for him we wish him the best.[238]

Council President Gary Goodheart was more forthcoming in his support for Hewitt. He said that he was not surprised on hearing of Hewitt's pending interview. Asked to elaborate on that, Goodheart commented that "Steve has a feeling the Council doesn't have a lot of confidence in him." Was this feeling unjustified? "No, I don't think so; I think he has legitimate concerns."[239] Goodheart's phone began "ringing off the hook" with people expressing their anxiety over the prospect of Hewitt's leaving. Said Goodheart:

> I don't blame them. Hewitt is Mr.Greensburg. I think most people know how vital he's been to this town's recovery.

Mark Anderson asked Goodheart if Hewitt would have been less inclined to consider leaving if the elections of April 2008 had been postponed a year. Goodheart replied, after a moment of reflection: "Let's just say I think there'd be different results right now."

In an editorial for the *Signal*, Anderson was unequivocal:

> Because of the streamlining and efficiency Hewitt brought to the City's management in the ten months he was at the helm before the storm hit, Greensburg was in far better position to even attempt, much less pull off a recovery from the meteorological mayhem it suffered that night.

. . . .

---

[237] staff, "Hewitt a candidate for Valley Center," *Pratt Tribune*, Jan. 19, 2009.
[238] Mark Anderson, "Goodheart hoping to encourage Hewitt to stay put," *KCS*, Jan. 19, 2009.
[239] *Ibid.*

You might . . . want to give the current mayor a call. Implying that Hewitt's interest in a higher profile, more gratifying position is more or less the natural evolution of events for someone who's enjoyed months and months of positive publicity for his work, and then saying you 'wish him the best' if he does happen to move on, is a less than effusive expression of appreciation.[240]

A blunt assessment of the reasons behind Hewitt's decision to apply for the Valley Center position came from John Janssen:

> It's not a great secret that there have been issues between Steve and the mayor. When I was mayor, I told Steve that he had been fighting this thing 24/7, he had all the contacts, and he knew what buttons to push when he needed something. . . . Bob Dixson is retired, he has nothing else to do, and he's been trying to micromanage Steve.

A Council meeting took place on the evening of January 20, 2009, in the same trailer where all Council meetings had been held since May 4, 2007. Not surprisingly, given the possibility of Steve Hewitt's leaving Greensburg, the trailer was filled to capacity. After disposing of more humdrum matters, the meeting got down to the serious business of the night. Several citizens spoke on Hewitt's behalf. These included a number of people who have played roles in this book: Farrell Allison, Marvin George, John Janssen, Gene West, Scott Brown, and Daniel Wallach. After their comments had been heard, the City Council retired to executive session, which removed it from public oversight. After an extension of the session, the Council returned to open session. One Council member, Brandon Hosheit, announced that the Council would do everything possible to retain Steve.

Interviewed in March 2009, Hewitt said that he was pleased to hear of his support in the community. The vote of confidence in the Council had given him the energy he needed to continue and "get the job done." The Monday following the Council meeting, January 26, 2009, Hewitt told his staff that he had decided to cancel his interview for the Valley Center position. "I don't think I really wanted to go," he said. "I have a new mind set of what this job in Greensburg can now be, and I think this is where I belong."

The idea of resigning his position and moving away from Greensburg was still in the back of Hewitt's mind. When asked in May 2009 if he was a tenant rather than a house-owner—as town gossip had it—he replied that he *was* renting but with an option to buy. "I know that, as City Administrator, that may be a problem, but I have a family to think of."

---

[240]Mark Anderson, *KCS*, Jan. 21, 2009.

Almost two years to the day that he admitted that he had been invited to apply for a position as City Manager of a Wichita suburb, Hewitt announced that he had accepted just such a position in a small town in Oklahoma. The *Signal* broke the news on January 5, 2011, in an article entitled "Comeback Kid not coming back."

"As a professional you look forward to new projects and new challenges," Hewitt told Mark Anderson. "That time for me was coming soon; I just didn't know it would come this soon." Off he went, the alpha male of the Planet Green series, the man whom Gary Goodheart had called Mr. Greensburg. But the years after the tornado—particularly 2009 to 2011—had taken their toll on Hewitt. He confessed in a late summer 2011 interview that he had lost both sleep and hair during his final years in Greensburg.

The Oklahoma town to which he was moving—Clinton—was one where Hewitt had previously worked as the parks and recreation director. In a follow-up article to appear in the *Signal* on February 5, 2011, Hewitt gave carefully diplomatic responses to the questions Anderson asked. There were no allusions to the nitpicking that bedeviled his job—issues like trash collection and the location of flagpoles in front on City Hall—which seemed at times to dim the glow of the Emerald City which was rising on the ruins of the old Greensburg. Had Hewitt grown tired of taking the heat on these issues? He did not say. When interviewed seven months later in Clinton, he was more frank in assessing the effect of the vitriol he had faced.[241] Reaction to his leaving was surprisingly muted after the pyrotechnics two years earlier, perhaps because Hewitt's resignation came as a *fait accompli*. Gary Goodman congratulated Hewitt on his move which, he said, would be good for Hewitt's whole family.

****

The other linchpin of Greensburg's sustainable development, Daniel Wallach, continued to bask in the public eye. *Kansas Country Living* featured Wallach on the cover of its March 2009 edition and ran an article about Greensburg GreenTown. The article preceding this one contained an interview with two state politicians, one of whom said that "Wind and other forms of energy are important but coal is necessary." Both politicians said that they would work to secure the votes to pass a bill allowing Sunflower Electric Power to build its two coal-fired power plants near the town of Holcomb. The interview went on to tout the importance of coal as a source of the country's energy.

---

[241] Hewitt complained of the stress of his job, particularly toward the end of his tenure as City Administrator of Greensburg. This led to a deterioration in his health.

Greensburg GreenTown fostered good relations with the townsfolk in a number of ways. At the time of the second anniversary, the organization sponsored a buffet banquet attended by 400 people. After a meal which featured locally raised bison burgers—"knowing where your food comes from and minimizing the distance it travels to reach your plate are traditional values that we aim to re-inspire"—Wallach presided over an awards ceremony, handing out 'Green Initiative Awards.' So many people received recognition that their names—with a brief commentary following each of them—filled four pages of the program. In fact, Wallach described the event as "kind of like the Academy Awards." First up was the National Renewable Energy Laboratory (NREL). There to receive the award on the agency's behalf was Lynn Billman, who let the audience know "what it had done:" Half of the 200 houses rebuilt or repaired in the previous two years were rated for energy efficiency, she said, and of those that agreed to be tested, 42 percent were more efficient than code. More recognition for her work would be coming Billman's way. In early 2011, she was notified that she would be receiving one of NREL's Outstanding Performance awards. Letters of support for Billman came from a number of Greensburg residents: Mayor Bob Dixson, Daniel Wallach ("Lynn Billman is awesome," all in caps), Steve Hewitt, Gene West, Mike Estes, Erica Goodman and Mary Sweet—a roll call of the movers and shakers of the Greensburg revival. "Creating change in the world for a better today. . .is pretty heady stuff," said the NREL awards presenter. "She changed the world by changing a town."[242]

Last up was Chuck Banks, "an entrepreneur trapped in a government official's body," according to Wallach.

There were two surprises: Mason Earles and Emily Schlickman, recipients of an eponymous annual award, received a parcel of land in Greensburg as a wedding gift! As planned, the two students continued their education after the year they spent in Greensburg. Emily Schickman entered the Graduate School of Design at Harvard in the fall of 2009 while Mason Earles enrolled in the Life Cycle Analysis program at the University of Maine. The distance between Bangor and Boston, Mason said, was not too daunting for the fresh fianceés. Emily and Mason got married in the spring of 2010. Their projected reception in Greensburg did not take place.

Lonnie McCollum returned to town to receive a Visionary Award named after him, just the way the Earles/Schlickman Award had been named after its first recipients. Attendees could only speculate how the former mayor felt about sharing the awards roster with the men whose criticism of him had ended his mayoralty. McCollum sat in one corner of the gym, close to his friends, Erica and Gary Goodman. The Estes brothers sat in another corner, far, far away.

---

[242] Mark Anderson, "Billman receives NREL award," *KCS*, Feb. 23, 2011.

Pressed later to reminisce on the changes in his life since the tornado, McCollum admitted to the feeling that there was a giant hole there, one that would probably never go away. He expressed disgust with politicians in general and said that he was considering a run for Congress. In a cascade of political dominoes, Jerry Moran was then running to replace Sam Brownback, the senior senator from Kansas, who was himself retiring to run for governor. McCollum thought that, by running for Moran's seat in the First Congressional District, he would be offering citizens a chance to vote 'no.'[243] McCollum had been working on his 'museum pieces,' the old 1932 Ford pickup, newly restored after the tornado, and a recent acquisition, a 1955 Chevy. The cars were a link to his childhood. Lonnie McCollum had had a happy childhood.

Jerry Diemart attended the banquet, but he would soon be yo-yo-ing back and forth to Texas to find work. There wasn't much of that in south-central Kansas. In a year, his mother would be managing a new motel— two stories and 42 rooms—again under the auspices of Best Western. The Nightwatch, as the new motel was called, went up three blocks west of the Business Incubator on US 54. It opened for business at the end of October 2010. Jerry briefly became the associate manager of the new facility, abandoning for a few months both his wanderings in search of gainful employment and his maintenance business in Greensburg. He also resigned from the board of Greensburg GreenTown. But by late 2010, he was back in the business of repairing furnaces and AC units.

****

Hewitt was not the only Greensburg man to have risen to greater eminence in the wake of the tornado. On November 18, 2008, Dennis McKinney announced that he had accepted an appointment by Governor Sebelius to fill the remaining two years of the State Treasurer's term, a position vacated by the previous treasurer when she won a seat in Congress. Despite the year-round responsibilities of his new position, McKinney opted to continue living in Greensburg. There was speculation that McKinney's political star was on the rise and that he might run for higher state office some time in the future. Unfortunately for him, the Republican sweep of the midterm elections in November 2010 cost him his job: despite support in the traditionally liberal northeast of Kansas and in Kiowa County—where he mustered 861 votes to his opponent's 195—McKinney lost, 42 percent to 58 percent. This defeat terminated nearly two decades of public service for one of the acknowledged heroes of the 2007 tornado.[244] Temporarily

---

[243] Both Moran and Brownback won their races in the Republican sweep of November 2010.

[244] Jim Schroth shed light on what had happened since the tornado to his daughter, Kelsey. This was the woman who had been rescued dramatically, along with her baby,

McKinney took over Hewitt's position while Greensburg sought a new City Administrator.

Three years after he denied a permit to Sunflower Electric Power Corporation to build a coal-fired power plant at Holcomb, Director of the Kansas Department of Health and Environment Rod Bremby was fired by Governor Mark Parkinson. Bremby's action had stunned the nation. Environmentalists throughout the country viewed his dismissal as an attempt by Parkinson to expedite the granting of a new permit to Sunflower to build the scaled-down plant which the governor had negotiated with the company shortly after assuming the governorship of the state. "There isn't anyone in the state who doesn't know what this was about," claimed Scott Allegrucci, Executive Director of the Great Plains Alliance for Clean Energy.[245] And guess what? On December 16, 2010, it was announced that Sunflower's permit was approved. Environmentalists were quick to respond. "Kansas gets the pollution," said Stephanie Cole of the Sierra Club, "Colorado gets the power."[246] Most of the new plant would be owned by Tri-State Generation, a Colorado-based company, and Transmission Association. It was reported that employees at the Kansas Department of Health and Environment, working on the permit, had put in time on weekends. "By turning the permitting process into a race against the clock, the state has signaled that it does not value public involvement," said Cole.[247]

With the change in administration on the horizon—Governor-elect Sam Brownback would take office in January 2011—Bremby would almost certainly have been ousted, anyway, so why the rush? Speculation focused on the stricter environmental laws which came into effect on January 1, 2011. These would have made the plant at Holcomb more expensive to build. Thus the need to move quickly. "A decision on a permit . . .was nearing Bremby's desk," noted one observer.[248] Even with Bremby out of the way, there were roadblocks on the way to constructing the power plant. A legal challenge was undertaken by the Sierra Club which alleged that it was necessary to do an environmental study of the project and to

---

Jayden, by McKinney. Schroth said that Kelsey and her boyfriend, Chris Koss, were still together. Kelsey had gone on in 2008 to obtain a license as a practical nurse (LPN). Jayden was growing up and spending some time with his Granddad. On one of their fishing trips together, a whistle went off in the distance, and Jayden told Schroth, "Tornado, Grandpa."

[245] Karen Dillon and David Klepper, "Firing improves chances for Sunflower coal plant," *Kansas City Star*, Nov. 4, 2010.

[246] Scott Rothschild, "Coal plant permit approved," *Lawrence Journal-World*, Dec. 17, 2010.

[247] *Ibid*. Bremby himself said of the review process that "there were abuses." He went on to urge people to "look at the lobbying dollars in this process; it's staggering." See Bob Sommer's article, "Enough is Enough: Sunflower's Permit Should Be Revoked," in the Aug./Sept. 2011 issue of *Planet Kansas*.

[248] Scott Rothschild, "Gov. defends coal-plant actions," *Lawrence Journal-World*, Nov. 9, 2010.

conduct public hearings. The federal judge who heard the case agreed with the Sierra Club, which said, through one of its lawyers, that the decision would certainly delay the Holcomb plant. It might even lead to its abandonment.[249]

Another man to leave the theater of operations was Tod Bunting. The adjutant general concluded his duties as head of the Kansas National Guard the same month—January 2011—that Sam Brownback became governor. In an interview with the Associated Press, Bunting singled out 2007 as the most challenging of his seven years of service with the National Guard. Besides the strain imposed on the Guard by the wars in Iraq and Afghanistan, he singled out the natural disasters which had befallen the state that year. These included, of course, the Greensburg tornado.

"We don't want any storms and we don't really want to be at war," he said. "We join to serve, we join to be of service, so use us."[250]

Quietly slipping away himself, Mark Anderson, editor of the *Kiowa County Signal* left his post in June 2011, to minister to a congregation in northern Kansas. In a valedictory appearing in the paper on June 22, 2011, Anderson wrote that "as a preacher I'm unapologetic for what I believe is the biblical perspective when commenting on the public scene—a thoroughly conservative viewpoint to be sure. . . . And shouldn't we all, whether clergy or lay, have the same commitment to promulgating the truth as revealed to us?" His leaving was the occasion of Alanna Goodman's return to Greensburg. Her boyfriend, the TV cameraman who had worked for *Pilgrim Films & Television*, became Anderson's successor as editor of the *Signal*.

****

FEMA's Dick Hainje had mentioned the possibility of partnering some US towns with counterparts in the People's Republic in October 2008. His discussion with the Chinese bore fruit, and from November 27 to December 3, 2008, Hainje led a delegation of Greensburg officials which included Bob Dixson, Steve Hewitt, and Darin Headrick to Mianzhu City in Sichuan Province, a town that was leveled by an earthquake on May 12, 2008. The eco-partnership which paired Greensburg and Mianzhu was one of seven arranged between US and Chinese cities. The Greensburg-Mianzhu partnership was projected to last three years. It would include a sharing of views on how best to respond to disasters like those that destroyed the sister cities. In mid-March 2009, Greensburg GreenTown announced that

---
[249] John Hanna, "Sierra Club cheered by ruling in coal fight," *Lawrence Journal-World*, April 1, 2011.
[250] John Milburn, "Tod Bunting discusses tenure as Kansas adjutant general," *Kansas City Star*, Dec. 13, 2010.

its first two guest editors would be Lin Yang and Minghua Li from Han Wang, a town which lies in the jurisdiction of Mianzhu. The two Chinese students, who were studying in Boston at the time, would do case studies of the eco-partnered towns. Lin Yang made available a copy of their recommendations for Han Wang. They urged that all public buildings be built to LEED certification and that the town develop its ecotourism potential. Their discussion of events in Greensburg included a no-holds-barred assessment of the local opposition to going green:

> At the town hall meetings, Wallach and Hewitt encountered strong skepticism from many local residents, who were preoccupied with just putting a roof above their heads. In this rural, heartland town, it took creative reframing of the concept of sustainability to win community support. . . . When residents learned that building sustainably could translate to a 40% savings in utility costs, they became more receptive.

At the end of October 2009, you could strap on a hard hat and tour the new Kiowa County Hospital with Mary Sweet. Completion date for the building was set for early 2010. The hospital in the fall of 2009 was certainly a work in progress. Electrical cables snaked across the floor. A Spanish-speaking crew on cherry-pickers painstakingly added a final lick of paint to high walls. There was a slurry of last-minute alterations to sort through. Placement of furniture and partitions were fussed over. "We need to keep an eye on the children in the room," Sweet told the foreman. "Noooo," cooed a passing worker who overheard the remark. "What would children be getting up to?" Mysteriously someone higher up the design ladder had decreed that a wall which had already been finished be gouged to accommodate three windows. Could that be finessed? And so it went. Good worries to have. Not the bad kind. The hospital was up and running in February 2010, and the grand opening took place on March 12 of that year.[251] At the dedication of the new hospital, Angee Morgan spoke of the seven months she had spent in Greensburg, living—as she told the audience— in a box. For her part, Mary Sweet reminisced about the bittersweet days immediately following the tornado when she and her staff lived rough. That included their use of portable toilets.

"We had to learn port-a-potty etiquette," she recalled, "which isn't so bad in summer, but on cold, windy days it can be challenging sitting and shivering in there, knowing a big gust can tip you over at any moment."

That was now a relic of the past. The new facility, in fact, aimed to be the first such LEED-certified hospital in the country. Bursting with local

---

[251] One amenity of the new hospital was a room specially designed for an obese—the medical term is 'bariatric'—patient. The bed could support a weight up to 1,000 pounds, and the toilet could give the Grand Canyon a run for its money.

pride, Mark Anderson gushed that "this hospital makes other facilities of the area look sterile, unimaginative and blandly stark by comparison." [252]

That fall the Greensburg elementary school was closed for several days due to an outbreak of swine flu, and so there were fewer students around than usual. A few months before, a wall in the new school complex had been toppled by the wind. It was a microburst, Darin Headrick said, with a five-minute gust clocked at 90-100 mph. This had knocked down the cinder block wall, all 120 feet of it. "But it fell in the best possible way," he added, smiling. The time it took to rebuild —three weeks—was made up to keep the project on track. Schools were scheduled to open in the fall of 2010. Headrick had kept his promise to start classes on time in the fall of 2007, after the tornado had wiped out everything but one of his buses. There was no doubt that the new platinum-certified complex would be ready to welcome students for the start of the fall semester, 2010.

In October 2009, Headrick had to meet with the school project contractors, McCownGordon. This weekly event brought together contractors, architects, and owners. The day's topic was the schools' wind turbine. It was to be installed in a couple of months. Six men participated in the meeting at the teleconference site across the street from the school complex, in one of McCownGordon's trailers. One of them was Joe Keal, BNIM's lead architect for the school reconstruction. He sat hunched over his laptop computer, typing away but contributing, from time to time, to the discussion. A voice emanating from a telephone speaker in the center of a work table fielded questions asked by men in the room. (Like the voice of the Wizard of Oz in his throne room?) Much of the conversation dealt with the adequacy of a 125 amp breaker. Headrick was concerned about the integration of the wind turbine, and the non-continuous power it generated, with the city's power supply. No problem, he was told. Net metering would be available. Greensburg would be the only town in Kansas, for the next twelve to fourteen months, where that would be the case.

The highlight of that autumn of 2009 was a meeting of the City Council in the new City Hall, the first such meeting to take place in the new facility. At the time of the second anniversary of the tornado, almost six months before, a tour of the City Hall had been scheduled, but that mission was scrubbed. Not enough progress had been made on the building. Now the civic hub of the town was ready for unveiling. The sleek new structure, with its triangular-panelled ceilings and its walls of recycled brick, was a beauty to behold. And, appropriately enough, BNIM had sent a delegation to show Council its reworked and down-sized plans for a Big Well Museum—plans which Council ultimately rejected. As for City Hall, some acoustical fine-tuning seemed to be called for: The Reverend Blackburn, who had delivered the invocation for the meeting, felt compelled to leave council chambers to

---

[252] Mark Anderson, "New hospital breath-taking," *KCS*, March 3, 2010.

shush a very subdued conversation taking place in the vestibule of the building. "We can hear every word you say in there," he admonished.

****

A week before the town celebrated the second anniversary of the tornado, there was a service at Lighthouse Worship Center, the Assembly of God church attended by Vernon Davis, the man President Bush had spoken to a few days after the tornado. The weather was threatening, but services proceeded much as usual in the windowless sanctuary. It was a day for men's ministry but with women joining in the musical part of the service—a big component of group worship with piano, electric guitar and organ to accompany the singers. Postmaster Randy Kelly built his sermon around a question which Vernon Davis had posed: How do people know that their sacrifice offerings are pleasing to God? "Vernon does not ask that question to seek an answer," Kelly told his audience. "He wants to get you thinking."

The service was followed by a lunch of casserole dishes, veggie and pasta salads, and pastries for dessert. Circular tables had been set up in the sanctuary to accommodate the diners. Rain squalls which had come and gone all morning long began to intensify. The sky grew inky black, and there were strokes of forked lightning. Pastor Christa Zapfe announced that the radio had just said that Greensburg was under a tornado watch. Lunch would have to be interrupted. The congregation was ordered to take refuge in the basement of the nearby parsonage. Congregants obediently shuffled out. Some brought their paper plates heaped with food. Those without umbrellas arrived at the parsonage, their clothes rain-streaked, their hair stringy and bedraggled. A fire broke out in a neighboring farmhouse, and that brought out a fire truck, its siren screaming in the midday gloom. Memories of that awful night two years before came flooding back. One young woman broke into sobs. A friend tried to console her.

In the basement of the parsonage, some people continued to munch away on their casseroles and pasta salad. Everyone sat on the floor in a circle, children playing with Pastor Christa's little pug, Toby, who trotted from one person to another, looking to be petted. Don't touch his face, children were warned. All around the room, people offered prayers and encouragement. Another young woman began to cry and leaned against the breast of an older woman for comfort. Pastor Christa kept offering assurances to buck up people's spirits.

After an hour or so, the watch was called off. Several of the congregation—including Vernon Davis and his wife—returned to the church to finish their lunch. People ate in near-total silence. The light was failing and everyone strained to hear the wind. The ever-present wind. Which could roar across the plains like the voice of an Old Testament god but which was now only whispering softly, a synapse of Vishnu and Shiva, playing hide-and-seek with itself amidst the houses and churches of the rebuilt town.

****

Can any or all of the Greensburg experiment be exported beyond Kiowa County? This question is being answered as this book goes to press. On April 27, 2011, an EF4 tornado tore through Tuscaloosa, AL, provoking loss of life and property. One month later, on May 22, 2011, another destructive tornado wiped out swaths of Joplin, MO.

Greensburg officials swung into action, organizing a three-day conference in early June 2011. It brought together civic leaders from Tuscaloosa and residents of Greensburg. The seed for the idea of this Recovery Workshop was planted by Matt Deighton, who contacted FEMA in May 2011. He was subsequently called by FEMA. The relief organization was on Tuscaloosa's case to get organized. Deighton said that "they [Tuscaloosa leaders] should get their rear ends up here because I figured we could show and tell them what we did, what worked and what didn't, warts and all."[253] FEMA's Steve Castaner attended this workshop as an observer.

What did the Alabamans hear from Greensburgers? Darin Headrick told the visitors "to think long and hard before rebuilding." Chuck Banks urged his audience to seek technical assistance from multiple sources—foundations, state and federal government—as Greensburg had done. Mary Sweet agreed, and—she added—"play the tornado card in your hand."

A second one-day workshop, this time organized by FEMA later in the summer of 2011, brought together representatives from six states and the District of Columbia. This was a FEMA show, with Steve Castaner as master of ceremonies. Mayor Dixson told this gathering that "We lost everything that day. The only sustainable thing we had was each other." Residents told their guests that if they needed any help, they could call on Greensburg.[254]

Greensburg GreenTown was not to be left behind. On October 27, 2011, Wallach's organization helped to launch GreenTown Joplin. It was announced that fall that money was being raised for a resource center in

---

[253] Mark Anderson, "Tuscaloosa leaders to visit Greensburg for Recovery 101," *KCS*, June 1, 2011.

[254] Matt Deighton became active in a drive to get area residents to buy gift cards for stricken families in Joplin. "Can you imagine walking up to a complete stranger and handing him/her a gift card," he was quoted in the August 24, 2011, edition of the *Signal*.

Joplin similar to the silo eco-home in Greensburg. And there were embryonic plans for a contest to develop designs for sustainable housing to go up right after a disaster struck.

One way to look at what occurred, 2007-2012, in this tiny Kansas town might be to consider Greensburg as a sort of pilot project for the entire country—Ben Jervey's opinion notwithstanding—and, in particular, the rural parts of the US which have had trouble retaining population and attracting young families to settle there.[255] But was the recovery effort in Greensburg a model only for rural America? This seemed to be on his mind when Mark Anderson was asked to participate in a "Talk of the Nation" show on NPR. To bring together newsmen from Tuscaloosa and Greensburg seemed logical enough, but "it wasn't all that practical," given the order-of-magnitude differences in the populations of Greensburg, on the one hand, and Tuscaloosa and Joplin on the other. There was another difference: Tuscaloosa and Joplin had sustained a lot of destruction, true enough, but they hadn't been wiped out as had Greensburg.

Significantly Tuscaloosa chose BNIM to develop a community rebuilding plan. Relying on his previous experience—including BNIM's work in Greensburg—Bob Berkebile urged upon the Tuscaloosa City Council the need for what he called "full, robust community dialogue." In a sign of the times, that dialogue was facilitated by social media, in particular MindMixer, a web-based platform which describes itself as a virtual town hall.[256] *Tuscaloosa Forward* solicited ideas to rebuild the city, and respondents offered up to 70 suggestions on subjects ranging from land use and urban design to streamline processes.

A platoon of green soldiers was molded by the Greensburg experience. The expertise they acquired could now be put to the service of other towns and counties. Every day brought more of Daniel Wallach's eco-tourists to town, to see for themselves what had happened on the Great Plains of the Midwest. And Greensburg GreenTown was collaborating on the production of a natural disaster recovery guide to show readers "what to do, what not to do, what works, what doesn't." Why reinvent the wheel when decisions to rebuild "with a sustainable future in mind" had to be made?

****

---

[255] In October 2009, a green coffee shop opened first on the ground floor of the Business Incubator. It then moved across the street to Scott Brown's strip mall. Proprietor Kari Kyle came back to Greensburg to visit family. She was so taken by the incubator that "the hair on the back of my neck stood up." She gave the matter some thought, talked it over with her husband, and decided to return to the town to sign a lease and set up shop—to do something she really believed in.

[256] See *townhall.tuscaloosaforward.com*.

Was it worth it? Going green? Well, . . .yeah, said Gary Goodman, for the sake of future generations. Mayor Bob Dixson agreed. And going green had given Greensburg "a brand," what Steve Hewitt called "a marketing dream." People were still talking about Greensburg, nearly five years after the tornado, observed Hewitt as he looked back over the five years he had spent in Greensburg as that city's administrator. We were true pioneers, he said. The Greensburg story was unique. Bob Dixson said that it had given the town an identity.

When asked about the lessons to be learned by other communities tempted to follow Greensburg's example, interviewees offered a variety of opinions which ranged from the general to the very specific. By sharing what it went through, Greensburg could save towns like Joplin a year in planning and executing their recovery. That was Gary Goodman's estimate. John Janssen thought that even if the Greensburg experience was not 100 percent transferable, there were conceptual lessons to be learned here. He agreed that educating city governments about what to expect, should disaster befall them, was an important plus. Steve Hewitt's opinion was that planning was critical. It might take time but it was worth it. Stick to your plan, even in the teeth of adversity. And use common sense, a point he hammered over and over again.

There was criticism of the dominant role which BNIM played, particularly in the immediate aftermath of the tornado, and what the Kansas City architects cost the city financially. Dixson advised that function, not just form, be addressed in designing the new structures.[257] Talk to the people who will live and work in those new buildings, he said. Some projects—he did not specify which—should have been thought through more carefully. In any event, don't make life-affecting decisions too fast. Some opportunities had been lost, according to John Janssen, because "certain people wanted their names on things." Some announcements had been premature. Planet Green did a lot of good insofar as it made available resources to the town. Also, the way it interviewed people "kept a lid on stupidity." Hewitt cited difficulties in getting people to bring their property up to code. He thought the city had bungled that effort.

The failure to bring jobs to town may not have been Greensburg's fault. Certainly the city tried hard enough to lure industry to Kiowa County. Greensburg was isolated, it had no previous history of industry, and there just was not enough incentive—schools and hospital notwithstanding—for industry to relocate in south-central Kansas. At least, the infrastructure to handle incoming firms was now in place, with the creation of an industrial park east of Greensburg. But the recession, which began a year after the tornado, had obviously put a damper on industrial development. By the

---

[257] Steve Hewitt agreed with this.

fall of 2011, the industrial park stood as void of industrial apparatus as the prairie which surrounded it.

But that might change. The German company, HIB Systems, whose wood-block product had been used in the construction of the Meadowlark House, expressed interest in building its US headquarters in Greensburg.[258] HIB's North American representative, Jan Hoetzel, said that the company was searching for investors who could sink $2 million into the venture. Greensburg was salivating to score after its roster of industrial losses. It offered HIB a 20-year lease on a five-acre lot in the new industrial park; a yearly rent of $1; free utility hookups; and provisions for potential tax abatements.

The prospective tenant looked more promising than its predecessors. On Jan. 17, 2012, a group of might-be investors—including Scott Brown and his wife, Susan—flew to Germany to see for themselves what HIB was doing at its Meissenheim factory. They stayed five days. Bob Dixson, one participant in this junket, said:

> I want to do everything we can to make sure our relationship with HIB is ongoing. We want them to have their factory here in Greensburg.[259]

Scott Brown was impressed by what he saw. "I have no doubt . . . that these are top-notch folks, and I think they've got a top-notch product." Still he expressed some caution about the campaign to snare the German company. "It looked like it was going to take a sizable but not insurmountable investment to get this thing in Greensburg and probably quite a bit of the money is going to come locally."[260] As this book was being packed off to the publisher, it appeared that HIB might indeed be coming to Greensburg. But it was not a done deal in late March, 2012. Signing off on such an accord would be important for two reasons: 1. It would finally bring long-sought-and-tantalizingly-elusive industry to town. 2. As part of the agreement, HIB would pay to finish construction of the Meadowlark House, which the company would then use as a local office.

Going green had gone well for Greensburg, particularly in the early years after the tornado. It kept us in the news, John Janssen observed, echoing what Steve Hewitt had said. People were more conscious of their impact on the environment; that was Mayor Dixson's opinion. Sustainability had worked as advocates had hoped. He added—pointedly—that this meant financial as well as environmental sustainability. It tied in with the town's heritage, after all.

---

[258] Patrick Clement, "Hoetzel: 'Our goal is to build our headquarters here in Greensburg'," *KCS*, Nov. 16, 2011.
[259] Patrick Clement, "Investors return from trip with optimism," *KCS*, Feb. 1, 2012.
[260] *Ibid.*

ACKNOWLEDGMENTS

Start with the residents of Greensburg. This book would not have come into existence without their cooperation. My acknowledgment of debt to them will be an omnibus; so many have worked with me, I feel that I am on a first-name basis with half the citizenry of Greensburg. Do I dare to single out any individuals? All residents were helpful, but I should cite, in particular, the assistance and the friendliness of the Gardiner and Goodman families. I may have been, in fact, Erica Goodman's first customer at the reopened antique store *Where'd Ya Get That*. All three mayors who served during the period 2007-2012—Lonny McCollum, John Janssen, and Bob Dixson—gave generously of their time. As did City Manager Steve Hewitt, Commissioner Gene West and his wife Jan. My thanks go to Lloyd Goossen and Gordon Unruh of the Mennonite community. And to Kenny Unruh, who outlined the work of Mennonite Disaster Service. *Danke, meine Herren.*

Meteorologists Mike Umscheid and Jennifer Stark sought to explain to me what a tornado is all about. The Mullinville volunteer firefighters and Greg Ellis, the fire chief of Coldwater, spoke to me about the role of first responders after a tornado has ravaged a community. I deeply appreciate their efforts.

Dick Hainje of FEMA's Region VII, graciously spent a couple of hours to detail the work of FEMA in coming to Greensburg's aid. His external affairs officer, Crystal Payton, labored assiduously to flesh out the background for his commentary. Adjutant General Tod Bunting and his director of public affairs, Sharon Watson, mirrored the FEMA effort on behalf of the Kansas National Guard. I acknowledge with gratitude the help of all four of these people. Lieutenant Colonel Tim Stevens of the Kansas Air National Guard spent an afternoon talking about the EMEDS which his unit erected in Greensburg. And Angee Morgan devoted another afternoon talking about what Kansas Emergency Management did to supplement the work of FEMA and the National Guard. Thanks, as well, to Lynn Billman who met with me twice to talk about the work of NREL in the Greensburg recovery. And to the ubiquitous Chuck Banks, of the US Department of Agriculture, who spoke enthusiastically on several occasions of his work and that of his agency.

Let me thank Melanie Klein of Kansas State for sharing with me the projects which her students and Todd Gabbard's did for the community. At the University of Kansas, assistance was provided by three faculty members: historian Bill Tuttle in exploring the context in which south-central Kansas developed; psychologist Chris Crandall in analyzing the way the *Kiowa County Signal* chose to present global warming; and an especial thank-you to architect Dan Rockhill and his students in Studio 804 for explaining and letting me observe the evolution of the 5.4.7 Arts Center project.

Steve Hardy and Joe Keal of BNIM met with me separately on several occasions to share their insights into what their firm did for Greensburg. For an anarchist's perspective on what happened in the aftermath of the tornado, I am deeply in debt to Joe Carr and Dave Strano. Analysis of the role of Greensburg GreenTown was generously provided by Mason Earles, Emily Schlickman, and Joah Bussert. I need to thank, as well, Rob Threlkeld of the Johnson Foundation for his lightning-fast responses to my e-mail queries about the Meadowlark House.

Aaron Barnhart, TV critic for the *Kansas City Star*, offered encouragement and insight, as did TV analyst Gavon Laessig. Thanks, guys. I am beholden as well to Kathryn Takis, executive producer for *Pilgrim Films & TV*, for her prompt and thoughtful answers to my questions.

Gracias también to Wes Jackson and Ken Warren of The Land Institute for their hospitality and their enlightening discussion of agricultural policy in this country and elsewhere.

To Lin Yang of the Harvard Kennedy School, an expression of gratitude for making available a copy of the paper which she co-authored with Minghua Li, "Rebuilding a Sustainable Han Wang." To Chris Delbridge of the Victoria Department of the Environment, a warm thank-you for hosting my wife and me during our visit Down Under. To Charles Reich, author of the seminal book, "The Greening of America," who listened tolerantly to my exposé of the greening of Kansas late one sunny morning in San Francisco.

Technical assistance with LaTeX, my software package of predilection, was offered first by Denis Popel and then by Justin Graham. Where would I have been without them? The same sentiments in spades to my friend, Terry Smith, who did a crackerjack job of copyediting the manuscript.

And last, but not last, to my wife, Jean Grant, who offered advice, criticism, and encouragement, through the years of researching and writing this book. *Merci, mon chou.*

Index

KCS, 163, 166, 168

Agriboard, 143, 189
Allison, Farrell, 24, 126, 161, 162, 208, 212
anarchists, 55, 57, 58, 62
Anderson, Mark, 26–28, 159, 170, 192, 202, 204, 208, 211, 217, 222
Arts Center, 132, 134, 135, 137, 138, 140
Australia, 180–182, 188

Banks, Chuck, 99, 102, 103, 130, 158, 214, 221
Barnes, Stacy, 131, 133, 203
Bassets, Marc, 189
Berkebile, Bob, 5, 95, 113, 204, 205, 222
Berry, Wendell, 174
Big Well, 23, 116, 202, 204–206, 219
Billman, Lynn, 106, 107, 112, 149, 195, 214
biodiesel, 103, 183
Black Saturday, 181
BNIM, 94, 95, 97, 98, 103, 113, 144, 160, 162, 192–194, 202–204, 219, 223
Bond, James, 77, 80, 109–111
Bremby, Rod, 144, 176, 177, 216
Brown family, 107, 108
Brown, Lester, 10, 11
Brown, Scott, 93, 94, 190, 192, 203, 212, 224
Build It Bigger: Rebuilding Greensburg, 126
Bunting, Tod, 65, 68, 70, 75, 115, 217

Bush, George, 52, 67, 72, 75, 76, 117, 154, 155
Business Incubator, 103, 150, 197
Butler, Rex, 60, 61, 143, 172, 205

Cannonball Green, 17
China, 199, 217
City Hall, 162, 219
Clear Air Act, 176
Climategate, 167, 169
Commons Building, 162, 163, 207
Copenhagen conference, 8, 165
Costa Rica, 199
Crandall, Chris, 170

Davis, Vernon, 75, 76, 220
Deighton, Matt, 17, 110, 111, 208, 221
Delbridge, Chris, 181
DiCaprio, Leonardo, 91, 103, 119, 120, 200
Diemart, Jerry, 30, 41, 53, 98, 107, 140, 215
Dixson, Bob, 28, 181, 182
  as mayor, 124, 125, 134, 143, 156, 187, 193, 196, 203, 206, 211, 221, 224
Dyson, Freeman, 14, 175

Eco-Homes, 141, 142, 184, 186
Einsel, Aaron, 59, 61, 130
EMEDS, 74
End of Nature, The, 7
End of Oil, The, 12
Enhanced Fujita Scale, 3, 29
EPA, 175

Estes family, 54, 88, 99, 123, 180, 194, 195

Farley, Josh, 174
Fedrizzi, Rick, 100, 113, 138
FEMA, 52, 65, 66, 68, 70–72, 82–84, 95, 101, 125, 130, 164, 202, 221
first responders, 49–52
Flowerdale, 181
Foster, Norman, 199, 204

Gardiner, Chris, 37, 40, 41, 43–46, 53, 63, 84, 94, 146
Gardiner, Megan, 30, 37–40, 43–46, 53, 55, 63, 147
George, Marvin, 6, 145, 151, 198, 212
glanders, 18
global warming hoax, 166
Goodman, Alanna, 30, 33–35, 98, 191
Goodman, Erica, 30, 191, 193, 203, 205
Goodman, Gary, 26–28, 30, 34, 89, 187, 208, 209, 213, 223
Goossen, Lloyd, 22, 32, 105
Gore, Al, 166, 170
Green Bible, 6
Greensburg GreenTown, 90, 98, 106, 140, 186, 214, 221

Hainje, Dick, 65, 68, 76, 82, 83, 85, 217
Hardy, Stephen, 95–97, 101, 104, 112, 149, 160, 193, 203
Headrick, Darin, 28, 43, 77, 88, 112, 155, 161, 219, 221
Hewitt, Steve, 27, 51, 53, 87–89, 91–94, 103, 112, 113, 117, 125, 130, 143, 155, 183, 189, 192, 204, 206, 209, 210, 212, 213, 223
HIB, 186
House Select Committee, 118, 119
Hurricane Katrina, 66, 67, 70, 75, 85, 117

HydroSwing, 138

ICF, 105, 198
Inhofe, James, 13, 167
IPCC, 6, 7, 168, 169

Jackson, Wes, 171–174
Janssen, John, 28, 49, 79, 125, 133, 145, 156, 157, 162, 164, 205, 206, 208, 212, 223
    as mayor, 89, 93, 96, 100, 104, 107, 122
Jeffers, Dave, 195
Jervey, Ben, 200
Jones family, 35–37, 50, 53, 70, 78, 109, 201
Jones, Phil, 167, 168
Joplin, 221, 222

K State, 129, 130, 142, 163, 185
KC Courthouse, 158
KC Memorial Hospital, 144, 146, 218
KCS, 19, 24, 27
KDEM, 71, 159
Keal, Joe, 160, 219
Koch brothers, 166
KPP, 196
Kretz, Steffen, 188
Kuryak, Timothy, 120–122
Kyoto Protocol, 165

Laessig, Gavon, 120, 121, 125
Land Institute, 171–174
Law Kingdon, 206
LEED, 3–5, 113, 118, 144, 158, 161, 162, 175, 194, 206
    platinum certification, 4, 96, 97, 101, 111, 112, 131, 135, 137, 138, 142, 145, 146, 185, 194, 197, 204
Lighthouse Worship, 75, 108, 220
Limits to Growth, 9
LTCR, 95, 129, 133, 158
Lunch Box, 30, 146, 201

Main Street, 104, 155, 189, 190, 202

Mann, Michael, 182
Manske & Associates, 101, 112
Masdar, 199, 200, 202, 204
Mass. v. EPA, 175
Master Plan, 92, 95, 97, 192–194, 202
McCollum, Lonnie, 25–28, 30, 33, 53, 71, 123, 150, 192, 214, 215
  as mayor, 47, 61, 66, 72, 76, 87–89
McCownGordon, 219
McKibben, Bill, 7–9
McKinney, Dennis, 31, 42, 43, 103, 144, 177, 180, 215, 216
MDS, 22, 79, 80
Meadowlark House, 186, 187
Melbourne, 180, 181
Mennonites, 20–22
Moffitt, David, 184
Moran, Jerry, 76, 95
Morgan, Angee, 66, 71–73, 197, 218
MVP, 97, 101, 103, 112, 162, 163, 190

National Guard, 51, 58, 66, 68, 70, 81, 115, 116, 158
Nolan, Harry, 157
NRA, 59
NREL, 106, 144, 160, 185, 195, 214

O'Neill, Eileen, 91, 119, 121
Obama, Barack, 183, 187, 188

Pachauri, Rajendra, 8, 168, 169
Parkinson, Mark, 173, 178, 179, 216
Patriot Guard, 151, 153, 155
Payton, Crystal, 70, 73, 76, 82
Picard, John, 92–95, 100, 101, 113, 114, 142
Pilgrim Films, 91, 119, 120
Piligian, Craig, 91, 103, 120, 197
Plan B, 10
Planet Green, 91–93, 96, 100, 108, 119, 121–126, 136, 137, 151, 184, 196, 197, 223
Prairie Festival, 173, 174

Rat Pack, 28, 89, 90, 112
Red October, 71, 76
Risch, Thomas and Koegler, Sebastien, 127
Roberts, Pat, 75, 76
Roberts, Paul, 12, 13
Robinett Building, 190, 191
Rockhill, Dan, 130, 132, 135, 136, 138, 139

Schlickman, Emily and Earles, Mason, 140–143, 214
school complex, 160, 219
SCKTRO, 77, 80, 102, 109–111
Sebelius, Kathleen, 52, 66, 76, 95, 116, 118, 157, 164, 177–179, 188
Sierra Club, 176, 177, 179, 216
Silo House, 142, 184, 186
SIP, 105, 161
Smith family, 109
Smith, Levi, 100, 122, 202
Soulek, Al, 107, 108
Spangenburg, Ron, 208, 209
Stevens, Tim, 74, 144
Studio 804, 130–137, 139, 149, 160
Sunflower Ammunition Plant, 135, 138
Sunflower Power, 176, 177, 179, 213, 216
Supreme Court, 175
Sweet, Mary, 73, 144, 146, 218, 221

Takis, Kathryn, 121, 123–126
TAR, 7
Threlkeld, Rob, 186
Tuscaloosa, 221, 222
Twilight Theatre, 23, 24, 51, 96, 161, 162, 207

Umscheid, Mike, 31, 32, 63, 76, 77
Unruh, Gordon, 180
Unruh, Kenny, 79–81
Unruh, Tobias, 21, 22
USDA, 102, 103, 118, 130, 146, 159, 164

USGBC, 3, 4, 100, 133, 160

Wallach, Daniel, 89, 90, 98, 99, 106, 107, 122, 124, 125, 141, 186, 194, 209, 212, 213
Warren, Bill, 162, 208
Warren, Ken, 171–173
Watson, Sharon, 68, 115, 116
Wedel, Ruth, 121, 204
West, Gene, 112, 158, 207, 212
West, Jan, 163
Westboro Baptist Church, 152, 153
Wetmore, Bob, 187
wind, 3, 15, 17, 29, 36, 41, 43, 93, 97, 112, 149, 196, 219, 220
wind farm, 195, 196

Xtreme Structures, 142, 184, 185

CPSIA information can be obtained at www.ICGtesting.com
Printed in the USA
LVOW080752121012

302423LV00002B/2/P